Mathematical Modelling from a Cognitive Perspective

This edited volume provides readers with a snapshot of some of the current thinking on mathematical modelling education. It focuses on the relationship between cognitive and metacognitive activities and student approaches to attempting to solve real-world modelling problems.

This book presents a focused review of some current thinking and research on mathematical modelling education focusing on the relationship between cognitive and metacognitive activities and secondary student approaches to attempting to solve real-world modelling problems. The results of the review highlight areas where theoretical constructs are well advanced and point to areas where development and empirical confirmation are required. Four empirical studies showcase both qualitative and quantitative studies, providing recent findings resulting from a cognitive analysis of student modelling activity in classroom settings or under test conditions. Solving of data-rich modelling tasks; using strategic knowledge about drawing during modelling activities; connecting metacognitive individual strategies, metacognitive group strategies, and modelling competencies; and the activation and nurturing of students' mathematical thinking during the solution of a design problem are considered.

This book will be of interest to researchers, teachers and teacher educators, and curriculum and educational policy writers. It was originally published as a special issue of *Mathematical Thinking and Learning.*

Stanislaw Schukajlow, **Gabriele Kaiser**, and **Gloria Ann Stillman** are active members of the international research community which focuses on researching mathematical modelling education viewing modelling as real-world problem-solving. They have extensively edited and contributed to international journals and books on the topic. The latter two editors are past presidents of the *International Community of Teachers of Mathematical Modelling and Applications (ICTMA).*

Mathematical Modelling from a Cognitive Perspective

Edited by
Stanislaw Schukajlow, Gabriele Kaiser and Gloria Ann Stillman

Routledge
Taylor & Francis Group
LONDON AND NEW YORK

First published 2024
by Routledge
4 Park Square, Milton Park, Abingdon, Oxon, OX14 4RN

and by Routledge
605 Third Avenue, New York, NY 10158

Routledge is an imprint of the Taylor & Francis Group, an informa business

British Library Cataloguing-in-Publication Data
A catalogue record for this book is available from the British Library

ISBN13: 978-1-041-04434-5 (hbk)
ISBN13: 978-1-041-04436-9 (pbk)
ISBN13: 978-1-003-62833-0 (ebk)

DOI: 10.4324/9781003628330

Typeset in Minion Pro
by codeMantra

Publisher's Note
The publisher accepts responsibility for any inconsistencies that may have arisen during the conversion of this book from journal articles to book chapters, namely the inclusion of journal terminology.

Contents

Citation Information

The chapters in this book were originally published in *Mathematical Thinking and Learning*, volume 25, issue 3 (2023). When citing this material, please use the original page numbering for each article, as follows:

Introduction
Modelling from a cognitive perspective: theoretical considerations and empirical contributions
Stanislaw Schukajlow, Gabriele Kaiser, and Gloria Ann Stillman
Mathematical Thinking and Learning, volume 25, issue 3 (2023), pp. 259–269

Chapter 1
Modelling the phenomenon versus modelling the data set
Gloria Ann Stillman and Jill P. Brown
Mathematical Thinking and Learning, volume 25, issue 3 (2023), pp. 270–295

Chapter 2
Does strategic knowledge matter? Effects of strategic knowledge about drawing on students' modelling competencies in the domain of geometry
Johanna Schoenherr (née Rellensmann), Stanislaw Schukajlow, Judith Blomberg and Claudia Leopold
Mathematical Thinking and Learning, volume 25, issue 3 (2023), pp. 296–316

Chapter 3
Metacognition in mathematical modelling: the connection between metacognitive individual strategies, metacognitive group strategies and modelling competencies
Katrin Vorhölter
Mathematical Thinking and Learning, volume 25, issue 3 (2023), pp. 317–334

Chapter 4
Mathematical thinking about systems – students modelling a biometrics identity verification system
Ana Margarida Baioa and Susana Carreira
Mathematical Thinking and Learning, volume 25, issue 3 (2023), pp. 335–360

Notes on Contributors

Ana Margarida Baioa is a mathematics teacher in middle and secondary education. She teaches at the Group of Schools D. Manuel I – Tavira, where she is the coordinator of the Mathematics Department. She has a master's degree in Mathematics Education from the University of Algarve, Portugal, and is currently a doctorate student at the Institute of Education of the University of Lisbon, Portugal. She is an occasional collaborator in the Research Unit (UIDEF) of the Institute of Education of the University of Lisbon, Portugal. Her research interests include students' mathematical competencies, especially related to modelling and applications in mathematics education, as well as interdisciplinary teaching, and outside-school science and mathematics.

Judith Blomberg is a trained teacher and teacher trainer with many years of experience in mathematics teaching. She worked as a doctoral fellow at the University of Münster, Germany, until 2019. Her research interests are mathematical modelling, visualization strategies, attitudes, and beliefs.

Jill P. Brown is Associate Professor in Mathematics Education on the Burwood campus at Deakin University, Geelong, Australia. Her expertise is in mathematics education and in particular researching technology-rich teaching and learning environments (TRTLEs); affordances; and the teaching, learning, and assessing of functions, real-world applications and mathematical modelling, and mathematical thinking. Her research focuses on the primary and secondary levels of schooling from both the student and teacher's perspectives. She has been an elected member of the International Executive Committee of the research group, the International Community of Teachers of Mathematical Modelling and Applications (ICTMA), since 2013.

Susana Carreira is Associate Professor at the Faculty of Sciences and Technology of the University of Algarve, Portugal, where she teaches mathematics and mathematics education to graduate and post-graduate students. She is invited Associate Professor at the Institute of Education of the University of Lisbon, Portugal, where she collaborates in the supervision of master and doctoral students in mathematics education. Her research activity has been on mathematical modelling and applications, technologies in mathematics teaching and learning, mathematical problem-solving, mathematical creativity, and affect in mathematics education. She has participated in several national funded research projects and was the principal investigator of the Problem@Web Project (2011–2014). Internationally she has developed projects in partnership with Brazil, where she also co-supervises master and doctoral students in mathematics education. She has been actively involved in several national and international conferences, including the International Conference on Teaching Modelling and Applications (ICTMA), the Congress of European Research in Mathematics Education (CERME), and the International Conference on Technology in Mathematics Teaching (ICTMT). She has published research articles in national and international journals and authored and edited books, particularly related to mathematical modelling, problem-solving, digital technologies in mathematics learning, and affect in mathematics education.

Gabriele Kaiser is Full Professor of Mathematics Education at the University of Hamburg, Germany. Her areas of research are empirical studies on teacher education and teachers' professionalism, modelling, and applications in school, international comparative studies, gender, and cultural aspects in mathematics education. From October 2010 until October 2016, she held the position as Vice Dean of the Faculty of Education, being responsible for research, promotion of young researchers, and international cooperation. Since 2005, she serves as Editor-in-Chief of *ZDM – Mathematics Education*, an SSCI journal. She was Convenor of the 13th International Congress on Mathematics Education (ICME-13), which took place in July 2016 at the University of Hamburg, Germany. In July 2019, she was elected as President of the ICMI-affiliated study group 'The International Community of Teachers of Mathematical Modelling and Applications' (ICTMA).

Claudia Leopold is Professor of Educational Psychology at the University of Fribourg, Switzerland. Her research interests include learning strategies, mental visualizations, and learning with multiple representations. Stanislaw Schukajlow and Claudia Leopold are responsible for the project: Visualizations in solving modelling problems, within the framework of which the study was conducted.

Johanna Schoenherr (née Rellensmann) is Assistant Professor at Paderborn University. She was a post-doctoral fellow at the University of Münster, Germany, since her graduation in 2018. Her current research interests are mathematical modelling and visualization strategies. She studied Secondary School Education at the University of Münster, Germany, and majored in Mathematics and Social Sciences. In 2021, she finished her teacher training at a German secondary school.

Stanislaw Schukajlow is Professor of Mathematics Education at the University of Münster, Germany. His research interests are mathematical modelling, visualization strategies, emotions, and motivation in the context of mathematics learning. Stanislaw Schukajlow was a leader of the Affect and Mathematical Teaching and Learning Group of the 11th Congress of European Research in Mathematics Education. From 2014 to 2018, he served as an elected member of International Committee of PME (International Group for the Psychology in Mathematics Education). He is the Editor-in-Chief of ZDM – Mathematics Education.

Gloria Ann Stillman is Associate Professor of Mathematics Education at the Australian Catholic University, Ballarat, Australia. Her expertise is in mathematics education and in particular researching metacognition and the teaching, learning, and assessing of real-world applications and mathematical modelling. Gloria Ann Stillman was President of the international research group the International Community of Teachers of Mathematical Modelling and Applications (ICTMA) from 2013 to 2019. She is Series Editor with Prof. Gabriele Kaiser on the Springer book series, *International Perspectives on the Teaching and Learning of Mathematical Modelling*.

Katrin Vorhölter is Professor of Mathematics Education and Elementary Mathematics at Technical University Braunschweig, Germany. She was a researcher in the Department of Mathematics Didactics at the Faculty of Education at the University of Hamburg, Germany. She completed her post-doctoral thesis in 2020 on the subject of metacognitive group strategies of students in mathematical modelling.

Introduction – Modelling from a Cognitive Perspective: Theoretical Considerations and Empirical Contributions

Stanislaw Schukajlow, Gabriele Kaiser, and Gloria Ann Stillman

ABSTRACT
Mathematical modeling and applications are an important part of curriculum and considered to be important for students' current and future lives. In this contribution, we focus on mathematical modeling from a cognitive prospective. Following embedding the cognitive perspective within the discourse of mathematical modeling, we describe some of the current thinking on modeling activities and review empirical results. Subsequently, new findings and implications for research offered by the contributions of the current special issue are described. The contributions of this special issue refer to the relation between cognitive and metacognitive modeling activities to (a) mathematical thinking during building a modeling example, (b) strategic knowledge within modeling activities, (c) solving of data-rich modeling problems and their status from a meta-perspective and (d) individual and social metacognition.

Introduction

Origins of mathematical modeling as applying mathematics for solving real-world problems go deep into the past and can be traced to the early age of humankind. Solving real-world problems was a vehicle for arithmetic (e.g., via trade) and geometry (e.g., via measurement of the cultivated area of the Nile) centuries before mathematics appeared as a distinct scientific discipline (Joseph, 2011). In school education, however, for a long time, modeling was not in the focus of the curriculum in the academic stream of school education. At the beginning of the 19th century, the Meraner reform in Germany and in other parts of Europe, initiated by Felix Klein (1907), pointed out the importance of solving real-world problems for school mathematics, which can be seen as a turning point in curriculum development. However, several decades passed before modeling made its way into the school curriculum of many Western countries (e.g., Germany, Australia, and the United States of America) and, in recent years, East Asian countries (e.g., China and Japan). Since that time, mathematical modeling has not only been of interest to mathematics educators and teachers; but also it has increasingly become the subject of many research studies.

In their overview of different research approaches and perspectives on mathematical modeling, Kaiser and Sriraman (2006) identified as one of the influential approaches to mathematical modeling, a cognitive perspective, which they described as a meta-perspective as the aim of this perspective is to analyze cognitive processes which take place during modeling processes in order to enhance understanding of them. Consequently, the aims of this cognitive perspective are (a) the analysis and understanding of cognitive processes (e.g., students' individual modeling routes, barriers or strategies) and (b) fostering mathematical thinking while mathematical modeling (Kaiser & Sriraman, 2006). As modeling from a cognitive perspective describes a meta-perspective on mathematical modeling in education, the cognitive perspective offers an overarching view on the cognitive processes during mathematical modeling (Kaiser & Sriraman, 2006). Overall, modeling from a cognitive modeling perspective focuses on cognitive and metacognitive activities of students as a theoretical lens. As this perspective emerged in recent decades, not many studies have reviewed the state-of-the-art of research about this topic. To contribute to this research gap, we present a synthesis of findings of research on the cognitive perspective on mathematical modeling and summarize recent empirical findings regarding cognitive and metacognitive modeling activities. In the penultimate section, the contributions of this special issue are described.

Description of mathematical modeling and review of modeling literature from a cognitive perspective

Demanding transfer processes between real world contexts and mathematics are the core of mathematical modeling. In the discourse of mathematical modeling, the real world refers not only to phenomena from the world around us, but also to all possible extra-mathematical phenomena from the world of imagination (e.g., from fantasy stories or fairy tales) or from the scientific world (e.g., genetics, theoretical physics, chemistry or computer science, amongst others) (Niss et al., 2007). In a broader sense, the term mathematical modeling as real-world problem solving captures various cognitive activities that have to be carried out to answer a question anchored in the real world by using mathematics. Modeling problems include vague or unspecified conditions which make the solution process demanding. It is, therefore, not possible to solve a modeling problem simply by applying a well-known mathematical procedure.

In the cognitive perspective, researchers differentiate between transitions and states in the descriptions of modeling processes, the so-called modeling cycles. While solving a modeling problem, an initial solution approach can be refined and the problem is solved under new assumptions. The notion of "cycle" underlines the cyclical nature of modeling. Transitions in modeling cycles include modeling activities. Modeling cycles are idealized descriptions of modeling processes aimed at emphasizing the main phases in the ongoing solution process. Modeling problems differ considerably in their demands, which has led to the inclusion of different activities required for the solution of modeling problems being identified. We would like to emphasize that our idealized descriptions of modeling processes should not be confounded with modeling routes, which describe actual students' activities within modeling activities (see Borromeo Ferri, 2007). Different enactments of the modeling cycle reflect these different demands. An analysis of modeling processes in modeling cycles revealed the importance of the following activities: (a) cognitive activities: Understanding, Simplifying, Mathematizing, Working mathematically, Interpreting and Validating and (b) Metacognitive activities. Depending on the modeling perspective or research aims taken, some activities might be more important than other activities while solving a modeling problem. Furthermore, several activities can be summarized as one joint activity or sub-activities can be distinguished within one activity. Overall, the modeling process starts with an understanding of the real situation; in some approaches (e.g., Blum & Leiss, 2007), the phase of constructing a situation model is included (i.e., a mental model of the situation described in the task). The situation model has to be simplified and leads to a real model. Many descriptions of modeling cycles (e.g., Galbraith & Stillman, 2006; Stillman, 2011) go directly from the real-world situation to the real-world model. In most enactments of modeling cycles, it is then expected modelers construct a mathematical model via mathematizing the real model and work mathematically to obtain mathematical results. Mathematical results have to be interpreted and validated with respect to the assumptions of the initial model(s) and the given situation. During modeling activities, metacognition is involved (i.e., planning, monitoring, and regulation of cognitive activities). Importantly, students do not necessarily perform one activity after another activity, but often switch between activities while solving modeling problems, which are described by modeling routes (Borromeo Ferri, 2007). We now reflect on these activities from a cognitive perspective on modeling.

Understanding

Results about understanding of the real situation described in the task stem, in the modeling discourse, from linguistic and psychological research. While reading the task, students construct a mental representation of the situation, described as a situation model (Kintsch & Greeno, 1985). The important characteristic of the situation model is its holistic structure that allows making inferences during problem solving. A case study by Galbraith and Stillman (2006) identified students' blockages while understanding the situation described in modeling problems. These blockages were confirmed in other qualitative analyses of students' modeling processes (e.g., Göksen-Zayim et al., 2021; Schukajlow, 2011). Studies have demonstrated repeatedly a positive relation between understanding by constructing a real model and successful solution of modeling problems (Krawitz et al., 2021; Leiss et al., 2019). Other studies have pointed out that students can spend nearly half of the working time on reading comprehension (e.g., Leiss et al., 2019). In the study by Leiss et al., the time on reading comprehension depended on the linguistic complexity of the task. Applying reading comprehension strategies (e.g., note taking) was suggested to improve understanding. However, neither reading strategy training, nor prompting understanding by using reading comprehension questions, showed positive effects on worded modeling problems in subsequent studies (Hagena et al., 2017; Krawitz et al., 2021). One possible reason for these unexpected results might be difficulties in successful fostering of an adequate understanding of the real situation within the students.

Simplifying

The constructed complex situation model has to be simplified in order to construct a real model. Geiger et al. (2021) note that simplifications are necessary "to reduce the complexity of the extra-mathematical domain being modelled or to make the mathematics tractable." Prior research has confirmed students' difficulties with simplifying (Frejd & Ärlebäck, 2011). Considerations about how simplifying occurs are sparse. Krawitz et al. (2018) differentiated between two steps within simplifying activity that were applied while solving modeling problems with missing information: (a) noticing missing information and (b) making realistic assumptions, including identifying and supplementing the missing quantities. Mathematical knowledge might already be activated during simplifying processes and its activation affects the construction of the real model. Mathematical knowledge typically comprises mathematical concepts and procedures that can be used for mathematizing and working mathematically. One reason for students' difficulties with simplifying is attributed to students' beliefs that they have to combine the given numbers by a known mathematical operation in order to obtain a solution to a mathematical problem (Reusser & Stebler, 1997). These beliefs may come from limited experiences with modeling problems in the classroom. Interestingly, the study by Krawitz et al. (2018) indicated that students in their study did not have many difficulties with making assumptions if no numerical information was given in the task. Recent studies have revealed the relation of computational thinking and group creativity to simplifying (Cheng, 2021; Jung & Lee, 2021). International comparative studies have indicated differences between simplifying activities in different countries. A case study by Hankeln (2020) reported that German students used stronger simplifications compared to French students during modeling. Furthermore, German students made assumptions more often and consequently demonstrated higher modeling performance than Taiwanese students when their mathematical knowledge was controlled for (Chang et al., 2020). Ludwig and Xu (2010) demonstrated that German students did not outperform students from China when mathematical knowledge was not controlled. These findings emphasize the importance of mathematical knowledge for modeling activities, which might compensate for difficulties in simplifying and other cognitive activities.

Mathematizing

The process of mathematizing implies mapping objects and their relations in the real world to their counterparts in the mathematical world; but "Why can some students set up a model and others not?" One reason might be differences in students' knowledge of the real world and in their mathematical knowledge. Students have to anticipate the direction of a possible model formulation. However, understanding of the real model is not sufficient. In addition, profound understanding of mathematical concepts is essential. Niss (2010, p. 55) theorized that mathematizing activity was "a result of a knowledge-based mathematical anticipation of what is needed" that is aimed at linking objects in the extra-mathematical world to mathematics. Importantly, successful mathematizing relies on understanding and simplifying. Thus, some researchers suggest implemented anticipation of mathematics during understanding and simplifying being a prerequisite of successful mathematizing (Niss, 2010). Evidence of this phenomenon has been found in empirical studies (e.g., Stillman & Brown, 2014). Furthermore, the German approach of fundamental mathematical concepts (e.g., in German "Grundvorstellungen") suggests the importance of rich conceptual understanding of mathematical operations (Vom Hofe & Blum, 2016). Even elementary mathematical operations, such as addition of natural numbers, require understanding that either (a) two quantities can be combined to become a third quantity ("putting together"); or (b) one quantity can be added to another quantity resulting in the increasing of the first quantity ("subjoining"); or (c) one change is added to another change resulting in a third ("change"). Analysis of the tasks used in tests, such as examinations in the Netherlands, have demonstrated that students often do not need to simplify or mathematize the problems to find a solution (Vos, 2013). As mathematizing activity is built upon other activities, its assessment often includes other activities. For example, in a study by Greefrath et al. (2018), it was pointed out that using GeoGebra software for solving modeling tasks and working with paper-and-pencil (4 x 45 minutes long) on the same modeling tasks did not differ in their effects on mathematizing assessed by written tests. This result is not unusual, given that every digital tool relies on its own mathematical system and requires time and effort to learn the difference in syntax and sequencing of commands that correspond to "by hand" methods (Clark-Wilson et al., 2020) as were used in the assessing of mathematizing. Furthermore, new perspectives on mathematizing activities are offered by mathematics trails that require undertaking measurements, performing estimations and mathematizing the relations between quantities outside the classroom (Buchholtz, 2021; Hartmann & Schukajlow, 2021; Ludwig & Jesberg, 2015).

Working mathematically

Working mathematically implies applying conceptual and procedural knowledge to the constructed mathematical model in order to receive mathematical results. Research on mathematical problem solving revealed various individual and task characteristics that affect the success and failure in performing this activity. Problem solving cognitive and metacognitive strategies (e.g., forwards/backwards strategy and monitoring), prior mathematical knowledge, and students' beliefs have been revealed as important to solve mathematical problems (Goos, 2002; Schoenfeld, 1983). In the context of modeling, the role of technology while working mathematically is of specific importance. Technology can help to perform difficult calculations and free up time for other modeling activities. That is why it is worth considering what technology is allowed, or even recommended, to be used in each specific case of modeling. Research on using technology indicates that familiarity with technology and confidence in its use are the prevalent reasons for the choice of technology by students and that technology is mostly used by students for working mathematically (Geiger et al., 2003). However, some studies have demonstrated that a use of technology can affect positively other modeling activities, such as understanding, validating and interpreting (e.g., Mousoulides, 2011; Stillman & Brown, this issue).

Interpreting

Interpreting is the process that focuses on the reverse direction of mathematizing and includes the contextualization of the mathematical results. Due to contextualization, the solution process is moving back from the mathematical to the extra-mathematical world. When students solve purely mathematical problems or "dressed up" word problems, their mathematical results are identical with the final result of the problem and interpreting in a real-world context is not needed. Students with limited experiences with modeling problems may, thus, skip interpreting and finish their solution after calculating mathematical results. The question of how the mathematical result contributes to the overall solution of the problem (e.g., the length of the side in a triangle) is essential for modeling. The process of interpretation includes considerations about (a) the meaning of the calculated results in the extra-mathematical world and (b) meaningful accuracy of results (e.g., how many decimal places). Some researchers see interpreting as a part of validating activity (Albarracín et al., 2019). The extent to which interpreting is a blockage while solving modeling problems depends on the problems offered to students. For example, Schukajlow (2011, p. 185) identified students' difficulties with interpreting in one of three modeling tasks that students worked on. While solving this task, students performed several arithmetic calculations to attain the final mathematical results. Because of the multi-step nature of the mathematics used during working mathematically, students were not sure, at the end of their calculation processes, what their results meant in the real world. Similarly, Galbraith and Stillman (2006) noted that students have difficulties in interpreting their results. Furthermore, students often do not consider what accuracy is meaningful and round the result to, say, two decimal places. The accuracy of results depends strongly on the problem context and could be related to simplifying. If only a rough estimation of the given quantities is possible, it is not meaningful to give several decimal places for the real result. In other studies, interpreting was assessed along with other activities as a part of modeling. For example, an observational study conducted in middle school has indicated that the modeling classroom culture affected the interpreting activity (Brady & Jung, 2021).

Validating

Validating includes verification of the result within the real model and the situation model. If the result does not make sense, this might be an indication of errors in the solution process or of usage of an inappropriate real model. In this case, the understanding, simplifying, mathematizing, calculating and interpreting activities have to be examined and if results cannot be improved, the modeling process has to be carried out another time. One possibility to validate the result is asking whether this result makes sense. In some cases, it is obvious that the result is not meaningful (e.g., if traveling between New York and Tokyo was calculated to take one minute). In other cases, it is more difficult to decide whether the result is possible or not and an estimation can help to make such a decision. We know from prior research that, even if it is obvious that the result is not realistic, students often ignore this in the mathematics classroom (Blum & Borromeo Ferri, 2009; Jankvist & Niss, 2020). These studies indicate that students did not validate their results. One explanation of this behavior is that it might be seen as the teacher's job to control the solution (Niss & Blum, 2020, p. 120). Another explanation is that students notice that the result is not meaningful and ignore it as they do not have an alternative solution or model that can help to solve the problem. To overcome these issues, constructing multiple solutions was suggested to be helpful to overcome students' difficulties in modeling (Achmetli et al., 2019). By applying an alternative solution method, students can validate whether the results of the calculation are appropriate. An empirical study with four engineering undergraduates by Czocher (2018) demonstrated that validating activity is important for modeling and that it interacts with other cognitive activities in modeling (e.g., interpreting). This result indicates the close relation between validating and metacognition which we address now.

Metacognition

Metacognition plays "a critical role in successful problem-solving and modeling" (English, 2016, p. 9). In a broader sense, metacognition refers to knowledge about one's cognitive processes (Flavell, 1979). Garofalo and Lester (1985, p. 164) noted that "cognition is involved in doing, whereas metacognition is involved in planning and choosing what to do and monitoring what is being done." Consequently, planning, monitoring and regulation are essential components of metacognition. Before this, Pólya (1945) included the metacognitive processes related to "making a plan" and "looking back" in the description of problem solving. Validating occurs simultaneously with metacognition, as examining mathematical results implies monitoring and regulation of the solution process. In an analysis of group work and interviews in a case study, Stillman and Galbraith (1998) confirmed the importance of metacognition for modeling and revealed that metacognitive activities appear in different phases of modeling. During interviews, students demonstrated knowledge of cognitive strategies (e.g., organizing the information by using diagrams), and knowledge of metacognitive strategies (planning, monitoring and verification). The use of cognitive and metacognitive knowledge was positively related to students' success. However, whereas qualitative analysis in case studies repeatedly confirmed the importance of metacognition (Czocher, 2018; Galbraith & Stillman, 2006; Schukajlow, 2011; Stillman & Galbraith, 1998), the results on the positive relationship between self-reported strategy use assessed by questionnaires and modeling assessed by standardized tests are inconsistent. These findings indicate problems regarding the assessment of strategy use by self-reported questionnaires. Whereas knowledge about strategies (e.g., knowledge about drawing) can be assessed by using questionnaires (Rellensmann et al., 2020), strategy use while solving modeling problems often requires assessment via observations of students' problem solving behavior or by analysis of written solutions (e.g., note taking, summarizing or making a drawing, Bräuer et al., 2021; Leiss et al., 2019). Overall, many studies have been performed in the area of strategies and metacognition in recent decades (Maaß, 2006; Stillman, 2011; Tanner & Jones, 1995; Vorhölter, 2019; Vorhölter & Kaiser, 2016). Prompting students to develop multiple solutions, for example, improved planning, monitoring and self-regulation via the number of developed solutions (Krug & Schukajlow, 2020). Scaffolding learning mathematical modeling via a solution plan that included strategic help (e.g., "Link your result to the task and check if it fits!") improved self-reported strategy use compared to a control group and enhanced modeling activities after a teaching intervention (Hankeln & Greefrath, 2020; Schukajlow et al., 2015). The analysis of students' responses have revealed strategies for proceeding, strategies for regulating, and strategies for evaluating as distinct factors on the individual and group level in studies reported by Vorhölter (2019) and Vorhölter and Kaiser (2016). Another area of research in metacognition within modeling from the perspectives of doing modeling and facilitating the modeling of others is anticipatory metacognition (Galbraith et al., 2017) where "the emphasis shifts towards reflection that points forward to actions not yet taken, that is, recognizing (noticing) possibilities of what 'might be'" (p. 84).

Contributions in this special issue

In this special issue, researchers have examined, in a variety of ways, the cognitive perspective on mathematical modeling. Cognitive and metacognitive activities and approaches fostering these activities have been undertaken and analyzed comprehensively to contribute to the research base on modeling under a cognitive perspective.

Baioa and Carreira (this issue) undertook an analysis of mathematical thinking during modeling whilst students conceptualized a biometric identity verification system. The study analyzed, among other activities, understanding of the system and of the models, simplifying (referred to as idealization) and validating of the models. A main finding is that the process of development of a mathematical model posed high demands on simplifying and was challenging for students. An important practical implication is the promising idea of building a prototype while solving a problem. Design and operation of the prototypes stimulate modeling activities and contribute to proficiency in modeling.

In an intervention study, Rellensmann et al. (this issue) investigated students' knowledge about drawing strategies, contributing to research on metacognitive activities. They analyzed the effects of training in strategic knowledge about situational and mathematical drawings on students' accuracy in modeling (at the task level) and on modeling performance across tasks (at the student level). The multi-level analysis demonstrated the indirect effects of the training on modeling via knowledge about drawing and drawing accuracy. The main theoretical implication is for metacognition. Accordingly, knowledge about a strategy is essential, but not a sufficient condition, for successful performing of modeling activities. A practical implication is that teachers should pay attention to both knowledge about a strategy and the accuracy of students' drawings, while prompting students to solve geometrical modeling problems.

Stillman and Brown (this issue) investigated how students solved modeling problems that are anchored in data-rich situations, aiming at gaining insight into how students develop meaning when modeling data-rich situations and the mathematical models produced. Most student teams in this study took a modeling the dataset approach but the degree they were aware of this differed from choosing to do so, and acknowledging this, to believing they were not doing this, but instead, modeling the phenomenon itself. There was evidence all cognitive activities mentioned above occurred in the study but incidence of simplifying, making assumptions, verifying and validating was less than ideal. The results of the study have theoretical implications for research on modeling activities in data-rich situations distinguishing modeling the phenomenon from modeling the data. A practical implication for classroom teachers is to take into account students' tendency to model a dataset rather than the phenomenon, when this type of problem is offered in class.

Vorhölter (this issue) presented an intervention study that aimed to clarify the roles of metacognitive individual strategies and metacognitive group strategies in modeling. This study elaborated on the concept of metacognition by differentiating it into individual and social components and linking metacognitive activities to cognitive activities by analyzing the effects of the training of metacognition on cognitive activities. The main results are (a) the increase of metacognitive strategies and (b) no effects of the development in metacognitive strategies on modeling activities. Vorhölter suggested that the impact of metacognitive activities might be visible on activities of the overall modeling process and not on single activities such as simplifying or mathematizing. A practical implication is the importance of discussing the use of metacognitive strategies in the classroom, because it might improve metacognitive group strategies (i.e.. evaluation).

Conclusion

In this article, we have analyzed mathematical modeling from a cognitive perspective. We described central phases of the modeling cycle and cognitive and metacognitive activities that are required for solving modeling problems. Although the field is already quite advanced concerning theoretical constructs, this survey and other contributing articles show that further developments and further theoretical work are needed, strongly connected to empirical studies. As the cognitive modeling perspective represents an overarching view on the cognitive processes during mathematical modeling, these results are relevant for other perspectives on mathematical modeling. Within the framework of the cognitive perspective, different approaches were evaluated that aimed to scaffold students' work on modeling. Some of these approaches addressed specific activities in modeling (e.g., reading comprehension) whilst others addressed modeling more holistically (e.g., metacognitive interventions or use of technology). We would like to emphasize that other possibilities beyond the cognitive perspective can also be successful, such as stimulating students' interest or self-efficacy in solving problems, or pointing out what variables are important for constructing mathematical models. The results of our review are relevant for researchers in modeling, as they summarize what we already know about modeling activities and what theoretical assumptions were not confirmed yet empirically. The studies that are presented in this review, and studies that are included in this special issue, evaluated teaching on modeling comprehensively. These results can be used by teachers fostering modeling in schools and by teacher educators to demonstrate possible learning environments for future teachers.

Disclosure statement

No potential conflict of interest was reported by the author(s).

ORCID

Stanislaw Schukajlow http://orcid.org/0000-0003-0519-5276
Gabriele Kaiser http://orcid.org/0000-0002-6239-0169
Gloria Ann Stillman http://orcid.org/0000-0002-0996-9120

References

Achmetli, K., Schukajlow, S., & Rakoczy, K. (2019). Multiple solutions to solve real-world problems and students' procedural and conceptual knowledge. *International Journal of Science and Mathematics Education*, *17*(8), 1605–1625. https://doi.org/10.1007/s10763-018-9936-5

Albarracín, L., Arleback, J., Civil, E., & Gorgorió, N. (2019). Extending modelling activity diagrams as a tool to characterise mathematical modelling processes. *The Mathematics Enthusiast*, *16*(1), 211–230. https://scholarworks.umt.edu/tme/vol16/iss1/10

Baioa, A. M., & Carreira, S. (this issue). Mathematical thinking about systems – students modelling a biometrics identity verification system. *Mathematical Thinking and Learning.*

Blum, W., & Borromeo Ferri, R. (2009). Mathematical modelling: Can it be taught and learnt? *Journal of Mathematical Modelling and Application*, *1*(1), 45–58.

Blum, W., & Leiss, D. (2007). How do students and teachers deal with mathematical modelling problems? The example sugarloaf and the DISUM project. In C. Haines, P. L. Galbraith, W. Blum, & S. Khan (Eds.), *Mathematical modelling (ICTMA 12): Education, engineering and economics* (pp. 222–231). Horwood.

Borromeo Ferri, R. (2007). Modelling problems from a cogniive perspective. In C. Haines, P. Galbraith, W. Blum, & S. Khan (Eds.), *Mathematical modelling (ICTMA12): Education, engineering and economics* (pp. 260–279). Horwood.

Brady, C., & Jung, H. (2021). Modeling presentations: Toward an assessment of emerging classroom cultures of modeling. *Educational Studies in Mathematics*. https://doi.org/10.1007/s10649-021-10056-x

Bräuer, V., Leiss, D., & Schukajlow, S. (2021). Skizzen zeichnen zu Modellierungsaufgaben – Eine Analyse themenspezifischer Differenzen einer Visualisierungsstrategie beim mathematischen Modellieren. *Journal Für Mathematikdidaktik, 42*(2), 491–523. https://doi.org/10.1007/s13138-021-00182-7

Buchholtz, N. (2021). Modelling and mobile learning with math trails. In F. K. S. Leung, G. Kaiser, & G. Stillman (Eds.), *Mathematical modelling education in east and west* (pp. 331–340). Springer.

Chang, Y.-P., Krawitz, J., Schukajlow, S., & Yang, K.-L. (2020). Comparing German and Taiwanese secondary school students' knowledge in solving mathematical modelling tasks requiring their assumptions. *ZDM - Mathematics Education, 52*(1), 59–72. https://doi.org/10.1007/s11858-019-01090-4

Cheng, A. K. (2021). Computational thinking and mathematical modelling. In F. K. S. Leung, G. Kaiser, G. Stillman, & K. L. Wong (Eds.), *Mathematical modelling education in east and west* (pp. 19–33). Springer.

Clark-Wilson, A., Robutti, O., & Thomas, M. (2020). Teaching with digital technology. *ZDM - Mathematics Education, 52*(7), 1223–1242. https://doi.org/10.1007/s11858-020-01196-0

Czocher, J. A. (2018). How does validating activity contribute to the modeling process? *Educational Studies in Mathematics, 99*(2), 137–159. https://doi.org/10.1007/s10649-018-9833-4

English, L. D. (2016). STEM education K-12: Perspectives on integration. *International Journal of STEM Education, 3*(1), 1–8. https://doi.org/10.1186/s40594-016-0036-1

Flavell, J. H. (1979). Metacognition and cognitive monitoring: A new area of cognitive developmental inquiry. *American Psychologist, 34*(10), 906–911. https://doi.org/10.1037/0003-066X.34.10.906

Frejd, P., & Ärlebäck, J. B. (2011). First results from a study investigating Swedish upper secondary students' mathematical modelling competencies. In G. Kaiser, W. Blum, R. Borromeo Ferri, & G. Stillman (Eds.), *Trends in teaching and learning of mathematical modelling* (pp. 407–416). Springer.

Galbraith, P. L., & Stillman, G. (2006). A framework for identifying student blockages during transitions in the modelling process. *ZDM, 38*(2), 143–162. https://doi.org/10.1007/BF02655886

Galbraith, P., Stillman, G. A., & Brown, J. P. (2017). The primacy of 'noticing': A key to successful modelling. In G. A. Stillman, W. Blum, & G. Kaiser (Eds.), *Mathematical modelling and applications: Crossing and reseaerching boundaries in mathemaics education* (pp. 83–94). Springer.

Garofalo, J., & Lester, F. K., Jr. (1985). Metacognition, cognitive monitoring, and mathematical performance. *Journal for Research in Mathematics Education, 16*(3), 163–176. https://doi.org/10.2307/748391

Geiger, V., Galbraith, P., Niss, M., & Delzoppo, C. (2021). Developing a task design and implementation framework for fostering mathematical modelling competencies. *Educational Studies in Mathematics*. https://doi.org/10.1007/s10649-021-10039-y

Geiger, V., Galbraith, P., Renshaw, P., & Goos, M. (2003). Choosing and using technology for secondary mathematical modelling tasks–choosing the right peg for the right hole. In Q.-X. Ye, W. Blum, K. Houston, & Q.-Y. Jiang (Eds.), *Mathematical modelling in education and culture* (pp. 126–140). Horwood.

Göksen-Zayim, S., Pik, D., Dekker, R., & van Boxtel, C. (2021). Mathematical modelling in Dutch lower secondary education: An explorative study zooming in on conceptualization. In F. K. S. Leung, G. Kaiser, G. Stillman, & K. L. Wong (Eds.), *Mathematical modelling education in east and west* (pp. 227–237). Springer.

Goos, M. (2002). Understanding metacognitive failure. *The Journal of Mathematical Behavior, 21*(3), 283–302. https://doi.org/10.1016/S0732-3123(02)00130-X

Greefrath, G., Hertleif, C., & Siller, H.-S. (2018). Mathematical modelling with digital tools—a quantitative study on mathematising with dynamic geometry software. *ZDM - Mathematics Education, 50*(1), 233–244. https://doi.org/10.1007/s11858-018-0924-6

Hagena, M., Leiss, D., & Schwippert, K. (2017). Using reading strategy training to foster students' mathematical modelling competencies: Results of a quasi-experimental control trial. *Eurasia Journal of Mathematics, Science and Technology Education, 13*(7b), 4057–4085. https://doi.org/10.12973/eurasia.2017.00803a

Hankeln, C., & Greefrath, G. (2020). Mathematische Modellierungskompetenz fördern durch Lösungsplan oder Dynamische Geometrie-Software? Empirische Ergebnisse aus dem LIMo-Projekt. *Journal Für Mathematik-Didaktik* 42 , 367–394 doi:10.1007/s13138-020-00178-9.

Hankeln, C. (2020). Mathematical modeling in Germany and France: A comparison of students' modeling processes. *Educational Studies in Mathematics, 103*(2), 209–229. https://doi.org/10.1007/s10649-019-09931-5

Hartmann, L.-M., & Schukajlow, S. (2021). Interest and emotions while solving real-world problems inside and outside the classroom. In F. K. S. Leung, G. Kaiser, G. Stillman, & K. L. Wong (Eds.), *Mathematical modelling education in east and west* (pp. 153–163). Springer.

Jankvist, U. T., & Niss, M. (2020). Upper secondary school students' difficulties with mathematical modelling. *International Journal of Mathematical Education in Science and Technology, 51*(4), 467–496. https://doi.org/10.1080/0020739X.2019.1587530

Joseph, G. G. (2011). *The crest of the peacock: Non-European roots of mathematics* (3rd ed.). Princeton University Press.

Jung, H.-Y., & Lee, K.-H. (2021). How mathematical modelling can be promoted by facilitating group creativity. In F. K. S. Leung, G. Kaiser, G. Stillman, & K. L. Wong (Eds.), *Mathematical modelling education in east and west* (pp. 377–387). Springer.

Kaiser, G., & Sriraman, B. (2006). A global survey of international perspectives on modelling in mathematics education. *ZDM - Mathematics Education*, *38*(3), 302–310. https://doi.org/10.1007/BF02652813

Kintsch, W., & Greeno, J. G. (1985). Understanding and solving word arithmetic problems. *Psychological Review*, *92*(1), 109–129. https://doi.org/10.1037/0033-295X.92.1.109

Klein, F. (1907). Meraner Lehrpläne für Mathematik. In F. Klein (Ed.), *Vorträge für den mathematischen Unterricht an den höheren Schulen* (pp. 208–220). Springer.

Krawitz, J., Chang, Y.-P., Yang, K.-L., & Schukajlow, S. (2021). The role of reading comprehension in mathematical modelling: Improving the construction of a real model and interest in Germany and Taiwan. *Educational Studies in Mathematics*. https://doi.org/10.1007/s10649-021-10058-9

Krawitz, J., Schukajlow, S., & Van Dooren, W. (2018). Unrealistic responses to realistic problems with missing information: What are important barriers? *Educational Psychology*, *38*(10), 1221–1238. https://doi.org/10.1080/01443410.2018.1502413

Krug, A., & Schukajlow, S. (2020). Entwicklung prozeduraler Metakognition und Selbstregulation durch den Einsatz multipler Lösungen zu Modellierungsaufgaben. *Journal Für Mathematik-Didaktik*, *41*, 423–458 doi:10.1007/s13138-019-00154-y.

Leiss, D., Plath, J., & Schwippert, K. (2019). Language and mathematics - key factors influencing the comprehension process in realitybased tasks. *Mathematical Thinking and Learning*, *21*(2), 131–153. https://doi.org/10.1080/10986065.2019.1570835

Ludwig, M., & Jesberg, J. (2015). Using mobile technology to provide outdoor modelling tasks-The MathCityMap-project. *Procedia-Social and Behavioral Sciences*, *191*, 2776–2781. https://doi.org/10.1016/j.sbspro.2015.04.517

Ludwig, M., & Xu, B. (2010). A comparative study of modelling competencies among Chinese and German students. *Journal Für Mathematik-Didaktik*, *31*(1), 77–97. https://doi.org/10.1007/s13138-010-0005-z

Maaß, K. (2006). What are modelling competencies? *ZDM*, *38*(2), 113–142. https://doi.org/10.1007/BF02655885

Mousoulides, N. G. (2011). Geogebra as a conceptual tool for modeling real world problems. In L. Bu & R. Schoen (Eds.), *Model-centered learning* (pp. 105–118). SensePublishers.

Niss, M., Blum, W., & Galbraith, P. L. (2007). Introduction. In W. Blum, P. L. Galbraith, H.-W. Henn, & M. Niss (Eds.), *Modelling and applications in mathematics education: The 14th ICMI study* (pp. 1–32). Springer.

Niss, M., & Blum, W. (2020). *The learning and teaching of mathematical modelling*. Routledge.

Niss, M. (2010). Modeling a crucial aspect of students' mathematical modeling. In R. Lesh, P. Galbraith, C. Haines, & A. Hurford (Eds.), *Modeling students' mathematical modeling competencies* (pp. 43–59). Springer.

Pólya, G. (1945). *How to solve it a new aspect of mathematical method*. Prinston University Press.

Rellensmann, J., Schukajlow, S., Blomberg, J., & Leopold, C. (this issue). Does strategic knowledge matter? Effects of strategic knowledge about drawing on students' modelling competencies in the domain of geometry. *Mathematical Thinking and Learning*.

Rellensmann, J., Schukajlow, S., & Leopold, C. (2020). Measuring and investigating strategic knowledge about drawing to solve geometry modelling problems. *ZDM - Mathematics Education*, *52*(1), 97–110. https://doi.org/10.1007/s11858-019-01085-1

Reusser, K., & Stebler, R. (1997). Every word problem has a solution. The social rationality of mathematical modelling at school. *Learning and Instruction*, *7*(4), 309–327. https://doi.org/10.1016/S0959-4752(97)00014-5

Schoenfeld, A. H. (1983). Beyond the purely cognitive: Belief systems, social cognitions, and metacognitions as driving forces in intellectual performance. *Cognitive Science*, *7*(4), 329–363. https://doi.org/10.1207/s15516709cog0704_3

Schukajlow, S., Kolter, J., & Blum, W. (2015). Scaffolding mathematical modelling with a solution plan. *ZDM - Mathematics Education*, *47*(7), 1241–1254. https://doi.org/10.1007/s11858-015-0707-2

Schukajlow, S. (2011). *Mathematisches Modellieren. Schwierigkeiten und Strategien von Lernenden als Bausteine einer lernprozessorientierten Didaktik der neuen Aufgabenkultur*. Waxmann.

Stillman, G., & Brown, J. P. (2014). Evidence of implemented anticipation in mathematising by beginning modellers. *Mathematics Education Research Journal*, *26*(4), 763–789. https://doi.org/10.1007/s13394-014-0119-6

Stillman, G., & Brown, J. (this issue). Modeling the phenomenon versus modeling the data set. *Mathematical Thinking and Learning*.

Stillman, G., & Galbraith, P. L. (1998). Applying mathematics with real world connections: Metacognitive characteristics of secondary students. *Educational Studies in Mathematics*, *36*(2), 157–195. https://doi.org/10.1023/A:1003246329257

Stillman, G. (2011). Applying metacognitive knowledge and strategies in applications and modelling tasks at secondary school. In G. Kaiser, W. Blum, R. Borromeo Ferri, & G. Stillman (Eds.), *Trends in teaching and learning of mathematical modelling* (pp. 165–180). Springer.

Tanner, H., & Jones, S. (1995). Developing metacognitive skills in mathematical modelling – A socio-contructivist interpretation. In C. Sloyer, W. Blum, & I. Huntley (Eds.), *Advances and perspectives in the teaching of mathematical modelling and applications* (pp. 61–70). Water Street Mathematics.

Vom Hofe, R., & Blum, W. (2016). "Grundvorstellungen" as a category of subject-matter didactics. *Journal Für Mathematik-Didaktik*, *37*(1), 225–254. https://doi.org/10.1007/s13138-016-0107-3

Vorhölter, K., & Kaiser, G. (2016). Theoretical and pedagogical considerations in promoting students' metacognitive modeling competencies. In A. R. McDuffie (Ed.), *Annual perspectives in mathematics education 2016: Mathematical modeling and modeling mathematics* (pp. 273–280). NCTM.

Vorhölter, K. (2019). Enhancing metacognitive group strategies for modelling. *ZDM - Mathematics Education*, *51*(4), 703–716. https://doi.org/10.1007/s11858-019-01055-7

Vorhölter, K. (this issue). Metacognition in mathematical modeling: The connection between metacognitive individual strategies, metacognitive group strategies and modeling competencies. *Mathematical Thinking and Learning*.

Vos, P. (2013). Assessments of modelling in mathematics examination papers: Ready-made models and reproductive mathematising. In G. Stillman, G. Kaiser, W. Blum, & J. Brown (Eds.), *Teaching mathematical modelling: Connecting to research and practice* (pp. 479–488). Springer.

Modelling the Phenomenon versus Modelling the Data Set

Gloria Ann Stillman and Jill P. Brown

ABSTRACT

This paper investigates students' mathematical modeling activity in data-rich modeling tasks. It aims at gaining insight into how students develop meaning when modeling data-rich situations and the mathematical models produced. A tendency to model a particular dataset, rather than the phenomenon that the dataset is a particular instance of, has been observed previously. Students concentrate on fitting mathematical objects such as functions to data, rather than using domain knowledge about the situation being modeled, mapping this to the data so as to capture the phenomenon as a whole. In other instances, students find functions that simply linearly interpolate the data and do not consider key features of the phenomenon, particularly when they have access to technological tools. The extent to which students' reasoning indicated awareness of their taking either approach was investigated in a qualitative study with Year 10/11 students. How the approach taken affected the processes students engage in whilst modeling was also investigated. The paper contributes to our currently limited literature on research into this issue and how it affects the outcome of students learning to model in classrooms at this level of schooling.

Introduction

Mathematical modeling is clearly part of mathematics and is now recognized as a fundamental mathematical practice for all students in many different educational jurisdictions worldwide (see, e.g.,, Barquero & Jessen, 2020; National Governors Association Center for Best Practices and the Council of Chief State School Officers, 2010; Villarreal et al., 2018; Vorhölter et al., 2014) and is increasingly seen as necessary in mathematics teacher education (e.g., Association of Mathematics Teacher Education, 2017; Borromeo Ferri, 2018; Gilbert & Suh, 2021; Villarreal et al., 2018). It is essential that teachers are able to select, implement, and assess modeling tasks in a meaningful manner that is commensurate with the hallmarks of mathematical modeling. To ensure they can do so, they need to know how students act when working with different types of modeling tasks. One such task type is data-rich modeling tasks.

This paper aims to investigate secondary school students' mathematical modeling activity in data-rich modeling tasks. In particular, it aims at gaining insight into how students develop meaning when modeling data-rich situations and the mathematical models produced. A tendency to model a particular dataset (e.g., the daylight hours forecast for a specific year at a particular location at 30-day intervals), rather than the phenomenon (e.g., the daylight hours for that location as they vary over time) for which the dataset is an instance, has been observed in the past (Galbraith, 2007) and similarly in sports contexts (Riede, 2003). We refer to these two approaches as *modeling the dataset* and *modeling the phenomenon*, respectively.

Students, and in some cases, their teachers (Riede, 2003), concentrate on fitting mathematical objects such as functions to data becoming a slave to the data rather than using domain knowledge about the situation being modeled (e.g., year length and timing of longest and shortest days of daylight) mapping this to the data so as to capture the phenomenon as a whole (Galbraith, 2007). In other instances, students find functions that simply linearly interpolate data and do not consider key features of the phenomenon (Berry, 2002; Maull & Berry, 2001; Zbiek, 1998) such as performance capacity of weight lifters not growing linearly as weight of the lifter increases as it becomes increasingly more difficult to perform necessary movements (Riede, 2003). It thus is appropriate to ask whether students are aware they are doing this and to investigate the effects these different approaches have on the modeling processes such students engage in when modeling these situations. This is particularly important in classrooms where students are learning to model and develop meta-knowledge about mathematical modeling, that is, what, why, when, and how they would mathematically model.

Mathematical modeling activity in data-rich tasks

Mathematical modeling

We agree with Caron (2018), who notes that mathematical modeling has to be an aspect of mathematical practice, which is an integral part of doing mathematics (see Stillman et al., 2020), whether that be as an engineer (Gainsburg, 2013), "a practitioner of mathematics, ... a mathematician, an economist, a biologist, any citizen, a teacher or a student" (Caron, 2018, p. 548). We also conceive of modeling as real-world problem-solving but acknowledge that it does not have to be. In this conception of mathematical modeling, understanding the situation being modeled and implications of the mathematical problem-solving for it, are of paramount importance to the modeling enterprise.

Mathematical modeling is a process, as many note (e.g., Caron, 2018; Doerr et al., 2017; Niss et al., 2007; Stohlmann et al., 2016), and can be represented as a cyclic activity (Gainsburg, 2013) that is iterative. A representation of this process as it proceeds from a messy real-world situation (A) to a series of stages (B-G), which are products of intervening actions related to cognitive activities, is presented in Figure 1. The thick arrows represent transitions in activity between stages with the articulation of specific cognitive activity (1–7) undergone by modelers between stages. These are often referred to as the *processes of modeling*. Such a representation is an idealized and simplified version of what occurs during modeling which, in reality, involves much more to-ing and fro-ing between stages. The process of mathematical modeling "is driven by considerations of both the external world and mathematics. The motivation for what to do next is a continuing give-and-take between the two" (Pollak, 1997, p. 101). The double headed arrows indicate reflective metacognitive activity that acts on the cognitive activities (Krüger et al., 2020; Vorhölter, 2018). This can involve looking forwards or backwards with respect to stages in the modeling (Stillman, 2011). The diagram itself can also act as a scaffold for beginning modelers that can be used by teachers during their interactions with students and by students themselves to monitor and evaluate their progress as was done in the study described here. For both

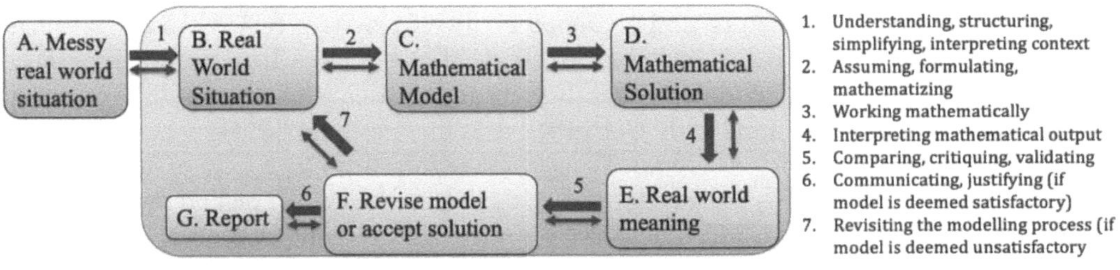

Figure 1. Representation of the mathematical modeling process (Stillman, 2011).

teachers facilitating the modeling and the students engaging in the modeling, such diagrams bring "an accompanying managerial understanding of when to appropriately use" (Yoon & Thompson, 2007, p. 204) particular cognitive activities and helps keep track of progress.

The mathematical modeling process as a whole builds "a mathematical model out of a real-life situation, and [uses] that model to analyze the situation, understand it, address an issue, solve a problem, make a decision, etc." (Caron, 2018, p. 548). Furthermore, any decision-making is not based on "portraying real-world argumentation as a unitary rationalistic phenomenon" (Gainsburg et al., 2016, p. 340) but rather, it is discipline or community of practice specific and thus must be related to the expectations and norms of the audience or community for whom the modeling was carried out, as well as the modeler's own convincing of self and others in the modeling team, during the modeling, that decisions are sound (Gainsburg et al., 2016). A modeling solution must be not only mathematically correct and explicable but also practical, giving reasonable and desirable answers in the real-world context (Pollak, 1997).

Empirical versus theoretical modeling

Mathematical modeling is a complex process and experienced modelers adopt different approaches depending on their perception of the real situation presented to them to model and the problem posed (Thompson & Yoon, 2007). Purposes of model building in such situations include, separately or in combination, measurement, deciding between alternatives, replication, prediction, explanation, or manipulation (Thompson & Yoon, 2007). Mathematical modeling can be characterized as empirical modeling or theoretical modeling (Berry, 2002) although a combination of the two can also be used. In an educational context, *empirical modeling* occurs when models are fitted to data that were either given with the problem statement or collected by the modelers, whereas *theoretical modeling* occurs when a model is constructed or developed from a theoretical perspective incorporating key features of the situation or phenomenon being modeled. In both approaches, data can be integral, however, because in theoretical modeling data can be used to verify the appropriateness of the model. Empirical modeling is employed when the purpose of the modeling is for prediction more so than understanding the underlying relationships of the phenomena involved. If the modeling is meant to give greater insight into a situation, then theoretical modeling is more appropriate.

Mathematical modeling tasks and approaches in secondary schooling

Bautista et al. (2014) characterize modeling activities or tasks as being on "a continuum along 'theory-driven' to 'data-driven' situations" (p. 9). The tasks discussed in this paper are data-rich tasks, which are considered to be toward the data-driven end of the continuum but that does not preclude students taking a theoretical approach to solution. However, the mathematical, digital display and analysis tools available influence both the mathematical structure perceived in a particular real-world situation and how mathematizing is attempted (Brown, 2015; Greefrath, 2020; Zbiek & Conner, 2006).

Zbiek (1998) previously conducted a study with a cohort of 13 prospective secondary school mathematics teachers who were novice modelers engaged in a series of data-rich, descriptive modeling activities constructing models directly from collected or provided data. They had access to curve-fitting software, graphing calculators, and a Computer Algebra System (CAS). Zbiek classified the participants' modeling approaches as: (a) Fitted Function Selector (FFS), where a curve fitter was used to generate all the software's fitted functions for the data and the one with best goodness-of-fit values was selected as the appropriate model; (b) Potential Function Generator (PFG), where a curve fitter was used to generate potential best fitting functions from the software's repertoire of functions but the appropriate model was selected by the match between the features of a function and those the modeler perceived as features of the real-world situation; (c) ScatterPlot/Graphing Tool (SCT), where scatter-plots of the data were produced but other functions beyond the tool's repertoire of prototypical functions to fit were graphed, and the model selected by a qualitative judgment of scatterplot fit and function characteristics; and (d) Unneeded/Unused Tool (UUT), where the modeler's judgments about relationships between relevant quantities in the situation without technological support were integrated withknowledge of mathematical entities such as ratios and functions to construct a model based on a small subset of the data. The first three approaches are in the empirical paradigm whilst the last is theoretically based and more toward theoretical modeling. According to Zbiek (1998), approaches were distinguished by the extent of modelers' reliance on the tool for model choice, the extent modeling was influenced by mathematical accuracy, mathematical understanding of relevance and/or real-world considerations.

The study of functions typically characterizes Year 9 – 12 curricula in many countries including Australia, especially in academic pre-tertiary mathematical pathways to university (Caron, 2018; Smith & Morgan, 2016). Perceived pressure not to take time away from mathematical content to be taught can result in compromises by teachers in jurisdictions where mathematical modeling has gained a foothold in curriculum documents and assessment requirements (e.g., Victorian Curriculum and Assessment Authority, 2015). According to Caron (2018), a Canadian tertiary mathematics educator, mathematical modeling in the last years of secondary schooling can often be reduced to finding the curve of best fit for a set of data points usually with the support of digital tools. This occurs in Australia as well. Such an approach to modeling, in Caron's opinion, limits experience of modeling processes relying on little development of understanding of the situation being modeled or gaining a deeper understanding of mathematics applied. However, Caron tempers her critique by noting the value in "using real life situations with their intrinsic complexity and authentic data" (p. 560). Agreeing with Maull and Berry (2001), she implores teachers to move away from the empirical paradigm, where manifestations of Zbiek's (1998) first three approaches to modeling using digital tools are likely to occur, to the theoretical paradigm.

What is it then that students are losing in this approach to data-rich modeling tasks and how might teachers counter this? A theoretical modeling approach requires an engagement with the underlying phenomenon at the level of understanding of the relationships between important variables in a situation. To achieve this, a structuring of the messy real-world situation (Figure 1) based on simplifications might allow the modelers to gain traction on the situation with the mathematical, representational, or analysis tools at their disposal. This can involve clarifying and employing physical, physiological, or social theories about the phenomenon or phenomena involved (see, e.g., Jensen et al., 2017). These can be based on researching the particular domain area where the problem resides, recalling experiential knowledge (Caron, 2018) or by consulting with experts in the domain (Vos, 2015). Modelers also form one or more mental models of the situation (Waisel et al., 2008), which can be refined by reflection on technology-generated external visualizations (Brown, 2015) as the modeling continues. However, to successfully model a situation there has to be an integration of the modeler's (a) physical/physiological/social theories about the relations between key variables of the phenomenon in the situation (Czocher & Hardison, 2021), and how these might change in relation to each other, if this is relevant to the situation being modeled; (b) the modeler's mental models of the situation; and (c) the mathematical modeling that the modeler employs. Having experience in being able to bridge gaps among these, and being able to successfully integrate all three aspects, is necessary if students are to develop a rich meta-knowledge about the modeling process as a whole. Teaching practices that lead to the curtailing of the necessity for such integration to structure a messy real-world situation will be a poor substitute for modeling in secondary classrooms. Using curve-fitting software and/or regression in computers or calculators, in a manner that allows modelers to go directly from data to a mathematical model, diminishes the modeling experience for them as they are not involved in decision-making related to the structuring and mathematizing of the situation being modeled, that is, the cognitive activities associated with transitions 1 and 2 in Figure 1, surrendering these to the technological tools they use. Another consideration, that might affect the modeling, is the possibility that some variables in the situation are not just controlled by physical/physiological principles but are subject to human decision-making related to factors external to the situation necessitating the testing of different scenarios, rather than just extrapolating from a best fitting function model.

In this paper, we seek to identify relationships between students' perceptions of the nature of their modeling approaches, as they work in small teams on data-rich modeling tasks, and their cognitive activities that affect their learning of mathematical modeling and meta-knowledge about the modeling process in a technology-enabled teaching/learning environment. In particular, we attempt to reveal the extent of their awareness of their modeling approaches leading to modeling the dataset, or part thereof, or modeling of the phenomenon and the perceived, versus actual, generality of their modeling.

Methods

To gain insight into this issue, we present a qualitative study where Year 10 and Year 11 students have been given data-rich modeling tasks in their regular classroom lessons as part of the *Using Mathematics to Solve Real-World Problems: The Role of Enablers Project* (DP17010555)[1] (Geiger et al., 2018). The research questions we will be addressing are as follows:

RQ1 (a) To what extent does student reasoning indicate awareness of modeling a phenomenon, rather than a specific instance of it, when upper secondary students (Years 10/11) are presented with data-rich modeling tasks? (b) Does access to technological tools influence prevalence of either approach?

RQ2 How does the approach taken (i.e. modeling the phenomenon or modeling the dataset) affect the processes students engage in whilst modeling?

Participants and their backgrounds

Thirteen students in a Year 10 class from one school, Heathcliff College,[2] in a one semester mathematics elective which had a problem-solving and modeling focus and 22 students in a Year 11 two-semester General Mathematics class at another school, Earnshaw College, participated. Both schools were Years 7–12 secondary schools located in outer and inner metropolitan areas, respectively, of a large Australian city.

The mathematical background of the Year 10 class included the study of linear functions for all students and non-linear functions for some students, plotting a dataset and finding a suitable model. The local state curriculum (Victorian Curriculum and Assessment Authority, 2016, p. 6) indicates it is expected that mathematics lessons will foster "the ability of students to make choices, interpret, formulate, model and investigate problem situations, select and use technological functions and communicate solutions effectively."

In Year 11 and 12, modeling is one of the essential mathematical activities described for all mathematics units (Victorian Curriculum and Assessment Authority, 2015) including Year 11 General Mathematics. Two aims of the senior curriculum show how "worthwhile and challenging mathematical learning" (p. 6) is intended to be implemented so that students can "apply mathematics to analyze, investigate and model a variety of contexts and solve practical and theoretical problems in situations that range from well-defined and familiar to open-ended and unfamiliar [aim 2] and use technology effectively as a tool for working mathematically [aim 3]" (p. 6). The Year 11 students' mathematical background included linear relations and equations, linear graphs and models, direct, inverse, and joint variation, investigating relationships in datasets between two numerical variables using scatterplots, Pearson's r and linear regression, and interpreting these in context, in addition to their Year 10 background.

The students also engaged in teacher-designed or selected real-world/realistic tasks during the semester and two video-recorded modeling sessions within the project. At Heathcliff College, these included the *Up Task*, about the lifting of a house with balloons based on the movie *Up* (Docter, 2009), as an introduction to the modeling cycle diagram in Figure 1 and its cognitive activities of modeling, and a version of the *Filling Up Task* (Blum & Leiß, 2006) before implementation of the first researched modeling task for this class. A further two data-rich tasks, the first comparing the salary spend by American baseball teams and their season wins, and the second comparing the careers of National Basketball Association of North America stars, Michael Jordan and Lebron James, were used before the implementation of the second modeling task for this class in the project. During a pre-lesson interview, the teacher described these tasks as involving "essentially just two sets of data plotting them against one another, and using the data to try and read it with functions on it."

The Year 11 students had engaged in only two other real-world open modeling tasks, *The Million Dollar Robbery* as an introduction to modeling and *Drought Proofing the Farm*, both conducted by their teacher. The modeling cycle (see Figure 1) was introduced during the introduction to the second task. This task was part of the data collection for the larger research project. Neither are data-rich tasks.

The tasks

Two tasks are the focus of this paper. The first, *Protective Service Officer (PSO) Task*, was about height estimation. The other task was related to sport as sport was a major interest of many in the class. Key details of the tasks are presented next.[3]

PSO task

The Minister for Police announced the recruitment of extra Protective Service Officers to improve public safety, especially during international events, but also to improve public confidence in using the transport network. Police Command has suggested that the PSOs are not good at estimating people's heights but are reasonably good estimators of small lengths, and where this is not the case, it can be easily remedied. The suggestion is that using a small body part length, easily estimated, a rule of thumb could then provide a good height estimate. Your team, as consultant mathematical modelers to Police Command, are to compile a report evaluating various rules of thumb that PSOs could use to provide a good estimate of height of offenders or suspects using shorter lengths. Police Command have provided some sample data. [Data supplied: first name, height, shoe length, knee height, wrist to elbow length, face height, gender, age, ethnicity for 41 adults aged 20–58, one with missing data.]

Weightlifting task (WLT)

In Olympic weightlifting one type of lift is the clean and jerk where the barbell is first lifted to the shoulders (the clean) and then, in a second motion (the jerk), to above the lifter's head. There are male and female categories and then weight classes within, so Women's 58 kg allows female lifters to compete who are up to this weight. Records are kept for each weight class. The world associations of weightlifting use a model that is based on the world records from the beginning of one Olympic year until the end of the next Olympic year. The International Weightlifting Federation (IWF) in 2018 disallowed all previous world records and started afresh. New records exist in some but not all categories. Your team, as modeling consultants to the IWF, is to provide 'world standards' where no record exists using your best models. The IWF will only ratify lifts as World Record Lifts that are 1 kg higher than the new world standards that also need to be both marketable and achievable. To increase public interest in weightlifting before the Olympics, a World Champion, irrespective of gender, will be crowned for different types of lifts by determining the best weightlifter across all weight categories. Your team is to argue who should be crowned World Champion for Clean and Jerk based on comparing the Actual World Record Lift with the predicted Performance Capacity of the lifter and to Rank the top 10 for a countdown. [Datasets showed current world record lifts for male and female weight classes with actual weight of lifter in super heavy class only, record holder's name and country, date record set as well as current world records for men and women with name of record holder and actual competition body weight in all classes.]

Table 1. Details of implementation of the two tasks at the two schools.

Task Implementation	Participants	Teams
PSO mid-August 2019 100 minute regular lesson at Heathcliff College	13 students in Year 10 elective class, 5 females and 8 males	5 teams of 2 and 1 of 3 – self allocated
PSO early September 2019 73 minute regular lesson at Earnshaw College	22 students in Year 11 General Mathematics class, all female	2 teams of 2 and 6 teams of 3 – teacher allocated
WLT late November 2019 100 minute regular lesson at Heathcliff College	12 students in Year 10 elective class, 5 females and 7 males	3 teams of 2 and 2 teams of 3–4 teams same self-allocated as *PSO*, fifth team formed by partner of an absent student combined with a pair

Note. All teams were single gender.

Data collection

Data for this paper were collected from implementations of the two modeling tasks. The *PSO Task* was used with the Year 10 students as their first video-recorded modeling task and with the Year 11 students as their second. The Year 10 students undertook the *WLT* as their second video-recorded task. In both tasks, students were provided with data, but could also collect more data themselves (via measurement for *PSO Task* or online for *WLT*). Students worked on the tasks in teams of two or three. One student from Heathcliff College was absent during the second task so there was one fewer team. See Table 1 for details of teams and implementations. Teams were either video or audio recorded as they engaged in the lessons led by their teacher.

Data used for analysis to answer the research questions came from scrutinizing the lesson implementation video-recordings of selected teams and their transcripts, together with transcripts of audio-recordings of the other teams, post-lesson focus group interviews (of 6–8 students from a mix of teams for each implementation), post-lesson video stimulated recall interviews with the teams video-recorded during implementation, and students' written reports of their modeling as they worked on the modeling tasks. In the stimulated recall interview, team members were asked direct questions about whether or not they considered themselves to be modeling the dataset given or modeling the phenomenon more generally at particular times during their solution, stimulated by short clips of the video-record of their team's working during the lesson. These questions were also posed to members of other teams in the focus group interviews.

Analysis

As a first pass through the data, the video files were viewed by both researchers, and all text data in the form of transcripts from video and audio files and the student written modeling reports were read and subject to data filtering by both researchers. Instances where the students' words or actions in images appeared to indicate that they were modeling the phenomenon, or restricting their approach to part or the whole dataset, or collecting data, were selected by each researcher separately. We then cross-checked that the selected images and passages of text or mathematical working were the same, discussing and culling or including until they were. An interim case summary (Miles et al., 2014) was produced for each of the groups and their approach to a specific modeling task. These were cross-checked by the researchers individually for completeness and ensuring that the context in which the modeling took place was preserved (Richards, 2005). To conduct thematic analysis, several qualitative, partially ordered meta-matrices (Miles et al., 2014) were constructed for data from the two sites. The meta-matrices assembled descriptive data from different cases in a standard condensed format (Miles et al., 2014) on the basis of inclusion of all relevant data to answer the research question in focus. Initially, these were quite large "mega-matrices" (Miles & Huberman, 1994) but they were gradually refined by the researchers transforming the case-level data into short quotations and summarizing phrases and then fracturing and clustering so they became more ordered and contained short, category-grounding phrases in each cell. This was an iterative process of data condensation and category refinement through a series of meta-matrices until emergent themes were distilled through the tactic of making comparisons and contrasting (Miles et al., 2014). At this stage of the analysis, exemplar cases were selected for the reporting of results.

Due to space limitations, only selected rows of case-level displays for partially ordered meta-matrices related to the first research question are shown from Heathcliff College (Table 2a) and Earnshaw College (Table 2b) but this analysis and categorization was done across all teams and tasks. As the tables show, the analytical categories of interest are the approaches taken by teams during modeling (i.e., modeling the dataset or modeling the phenomenon), evidence of awareness of the approach they were taking, the purpose they set for their modeling, and the alignment of their personal physical theories, mental models or images, and their mathematical modeling.

To understand the possible influence of digital tools (e.g., CAS calculators and laptop computers with Excel spreadsheets) on student approaches to the tasks and the processes engaged in, all data were coded for how digital technology was used (e.g., to make a scatterplot), modeling purpose of use (e.g., to verify mathematical output) using Villarreal et al. (2018, p. 333) codes, and modeling approaches of technology use when curve-fitting and regression software were available using Zbiek (1998, p. 198) codes. Additional codes were added as necessary and all coding was subjected to consensual agreement between the two researchers. See Table 3.

Finally, to complete answering the second research question, partially ordered qualitative meta-matrices showing the approach taken (i.e., modeling the dataset or modeling the phenomenon), and the modeling processes engaged in by students at the two sites were constructed. Each set of data for a group was coded for instances of cognitive activities in the transitions in the modeling process shown in Figure 1. Any additional cognitive activities related to modeling processes (e.g., identifying variables) were also coded. These codes were assigned individually by the researchers and then any discrepancies were discussed until consensus was reached. See Tables 4a and 4b.

In the coming sections, findings from this analysis will be reported, followed by a discussion in light of the literature and the research questions, and then conclusions will be drawn and possible implications for teaching and future research will be identified.

Findings

Selected exemplars from the meta-matrices addressing the themes that were distilled from the data related to students' approaches to the data-rich modeling tasks and their awareness of their approaches are presented to answer RQ1a. Possible influences of the availability of digital technology use on the student data-modeling approaches to the modeling tasks follow to answer RQ1b. Finally, the modeling processes engaged in by students during their modeling of these tasks are presented together with the modeling purposes of use of technology from Table 3 to answer RQ2.

Modeling part of dataset not the phenomenon – fully aware

If the dataset is perceived to be a set of numerical values where contextual meaning is secondary, other purposes than coming to a satisfactory resolution of the real problem presented in a modeling situation can be served quite readily. In the *WLT*, Ned from White Team (Heathcliff College) initially explored ratios and then tried to develop a linear model by hand. After some difficulty with finding a gradient, he decided completing the task with a report in the timeframe was his goal, rather than modeling the weightlifting situation in a realistic manner. He was totally aware in the post task interview that he was not modeling the phenomenon involved, as this response shows:

Table 2a. Selected rows from partially ordered meta-matrix: modeling approaches (Heathcliff college data).

Team	Approach	Awareness	Purpose	Alignment of Physical Theories/ Mental models or images/ Mathematical Modeling
PSO Task				
Green (Una & Tegan)	Modeling part of dataset – models of shoe length and knee height produced from *ratios of averages* based on data from 7 people in given data set – says shoe length "best" based on 5 trials from data of people in given data set.	Believe *model applies more generally in adult population* "because they are people" (Tegan, VSR). "Maybe not for children because children are a bit different. The growth rate is much different from ours. I don't mean ours, from adults" (Una, VSR)	Finding a rule that would allow PSOs to *predict* the heights of the people more accurately (Tegan, Report).	Physical Theory: "Shoe length is it's kind of, really it's better to the person's overall height, like the bigger shoe is obviously taller because you need bigger shoes to help balance the person more as they weigh more" (Una, VSR). Alignment between Physical Theory and Mathematical Modeling.
Blue (Eric & Ray)	At start just thought of data only as "numerical values" (VSR) then thought about them as lengths on their own bodies (Ray, Imp & VSR). Make 5 models for overall height based on *summation of all measurements, summation of multiples of selected combination of measurements* (so a proportional model) and use of *ratios of single measurements* (wrist to elbow and face height). They eventually use *averaging with proportional combination* model to generate Ht = Av FH + 2 x (Av KH + Av WtE) based on 40 in the dataset. Earlier models based on a few pieces of data ("focused more on those closest to the top") rejected on basis of lack of precision. Did not think to test on anyone outside of dataset (VSR).	Expected to be able to find a *perfect formula for the data set* (Ray, VSR). Eric says they are assuming that the measurements hadn't changed ("they have not grown since the measurements were taken") but then Ray says it wouldn't matter because they "would all be taken at the same time, regardless." (Imp)	Finding a perfect model to *predict* the heights of the people in the dataset (Ray, VSR).	Mental Image: picturing lengths mapped to make up person's height ("I kind of just looked at my own body and tried to figure out any patterns I could") – rejected shoe length as "a horizontal measurement, not a vertical one" (Ray, VSR & Imp); expecting "we could find something that was perfect" (Ray, VSR). Physiological Theory: Gender, ethnicity. age dismissed as "not a direct measurement, like something that would contribute to height, in my mind" (Ray, VSR). In some of the modeling, alignment between Mental Image, Physiological Theory & Mathematical Modeling.

(Continued)

Table 2a. (Continued).

Team	Approach	Awareness	Purpose	Alignment of Physical Theories/ Mental models or images/ Mathematical Modeling
White (Ned & Joe)	Modeling part of the dataset (36). Looking for a correlation between their height and measurements given. Looking for a pattern (Ned, Imp). Using four "imaginary people that have the averages of those" (Ned, Imp) gender x ethnicity. Hand sort and enter into Excel. Find *averages* for all "mini measurements" for shoe length, knee height, wrist to elbow, face height (including age) then *ratios* from height/ mini measurement (not age) producing *a table of multipliers*. "PSOs are now able to identify the criminals height by multiplying their measurement by the numbers provided categorized by their nationality and gender" (Ned, report).	Believes use of model is unrealistic method "'cause I don't like look at someone and then figure out how long or short their arms are. So I don't think anyone else would've." (Ned, FG)	*Prediction* – making a height "estimative tool" for PSOs "to be able to identify the height of people and suspects" (Ned, Report & Imp).	Physical theories: "If there was someone really young they could still have time to grow. But then I realized that each of the ages were up to 20 and that's usually when people stop growing." "And then the gender, I know that from science, that biologically women usually are smaller." Sister same height but shoe size "drastically different." Friends different races and ethnicity – different heights but trends "my friends who are more Asian based are usually smaller. Our shoe sizes don't differ that much." Age irrelevant as you "can't grow after the age of 26" (Ned, Imp). Alignment between Physical Theories and Mathematical Modeling.
Weight Lifting Task				

(Continued)

Table 2a. (Continued).

Team	Approach	Awareness	Purpose	Alignment of Physical Theories/ Mental models or images/ Mathematical Modeling
White (Ned & Joe)	Modeling part of dataset – ignore super heavy weight category "because it was so much more drastically different" (Ned, FG) and the effect it had was negligible. He did not think at the time to test the effect of this (Ned, FG). Ned anticipated there will be a *relationship* between "how much they weigh and how much they can lift" (Imp). He suggested converting the data into "a ratio so it is like for each kg they weight more than 55, . . . the weight they can lift increases by a set amount." By looking at *differences and averaging* by hand, he states: "For each 6 kg increase in weight [above 55 kg] they weight [sic can lift] 9.25 kg more" for male data (Ned, Imp). Wanting to continue a ratio approach by hand himself, he asked Joe to use Excel to produce *linear models*. "We put the data onto the computer, made a graph of the data and then we formed a trend line from the data" (Ned, Report). *Choice* of model *based on mathematical expediency as linear is considered easiest and quickest* (Ned, VSR). Ned uses his ratios as gradients in linear models which are very close to the computer generated versions. Uses computer models to generate missing world record lifts, compares predictions to actual lifts to select a World champion based on the lifters whose points were above the trendline being good performers, distinguishing between these by size of difference between actual & predicted lifts.	*Fully aware restricting to dataset given.* "I wasn't really having a model of humans doing weight lifting, I was just having a model, like based on the averages of numbers, to estimate this number [points to world standard needed]" (Ned, VSR). Both totally aware in post task interview they were not modeling the phenomenon. "Obviously, there is a limit to what humans can hold because if there wasn't, if any human can carry any weight then it wouldn't need robots" . . . "There probably was a limit and we probably should have taken that into account. . . . quickest way to have a linear" (Ned, VSR). Reason super heavyweight class omitted trying to make a linear graph & using point would have "curved it more" and "messed up everything." Claims better to use linear model with small dataset (Ned, FG).	*Prediction* – "you need to make a model to *predict* this [pointing to World Standards needed] (Ned, Imp). "Filling the empty parts of the table" (Ned, Report). "I feel like I kind of narrowed myself down just by viewing it as the whole goal was to find these two [world standards] (Ned, VSR). Seeking "a trend between how much they weigh and how much they can lift" (Ned Report).	Physiological Theories: Humans are limited by their physiology in how much they can lift and it would not "go forever" so it was acceptable to leave out the upper end (Ned, FG). Females have different physiology so expect their relationship could be different from that for males (Both, Imp). Misalignment between Physiological Theories and their Mathematical Modeling.

Note. FG focus group; Imp lesson implementation; VSR Video stimulated recall; FH face height, KH knee height, Ht height; SL shoe length; WtE wrist to elbow length.

Table 2b. Selected rows from partially ordered meta-matrix: modeling approaches (Earnshaw college data) to *PSO task.*

Team	Approach	Awareness	Purpose	Alignment Physical Theories/ Mental models/ Mathematical Modeling
Silver (Deja, Eve)	Modeling *full data set*. Write problem in form that can be mathematized: 'the PSOs are struggling to be able to estimate the height of people potentially envolving [sic] crimes.' Use Knee height versus height and wrist to elbow versus height with *scatterplot and Lin Reg and record only a, b, r and r^2* but seem unaware of how this could be used to predict height. Then did some *nonsensical calculations (height – shoe (max)) saying they are all means, min. and max* (Silver Imp, Report).	Appear to *think* it is an *artificial set of data* that they are modeling – "These people, they are real? Isn't Ken the Barbie doll?" "Who would name their kid Guy?" (Silver, Imp)	*Prediction* – "trying to find the height" (Eve, Imp). Deja struggles to understand purpose given height of dataset is known (Imp). (Mentions this throughout).	Physical Theory: "Like roughly on the age, kind of they are the same height because people are roughly around the same. It was like the same age, kind of like the same race and gender, were roughly the same height." (Imp). Flawed mental model of situation: "I did . . . maximum height take away the maximum shoe height to find the average" (Deja). "It gave you like a knee length average of people's height" (Deja). Misalignment of some of their modeling & mental model.
Pink (Iona, Jen, Kay)	Modeling part of dataset (40 – omit Ken). Need multiple rules as might be only able to get a specific estimate at a given time so no point just saying 'Shoe length is 'best.' Consider finding rules for a particular subset (e.g., males or Asian) to provide a more accurate rule if known, "if the regression was stronger when we only did boys then you could be more specific" (Kay, Imp). *Generate several linear models for several variables stating reliability using goodness of fit values*, e.g., height = 871.3707 + 1.6260894 *knee height. R = 0.7090863, "Oh, that is pretty strong" (Kay, Imp). Test one model on Alice & another on Bob. Report includes only values, although during modeling Kay notes "I just tested knee height [male specific model] on Bob and it works." Use r and r^2 to compare strength, report only list of 5 models with no discrimination, to impractical degree of accuracy, but stating in VSR would round in practice "because they are on the spot, on the job."	Iona and Kay say *realized* it was just a *specific set of data they were modeling* (VSR). In FG, Kay argues against rest of FG saying *solution to problem is realistic* & includes estimating from CCTV footage. In Imp realize they *need to test their models beyond data set* – on themselves if have time. "It will be interesting when we get our lines of best fit and test it on each other" (Kay). "Yes, see if it works" (Iona). "Yeah, see if it gives our heights" (Kay). "We tested it but the person we tested it on, was included in our making that model" (Kay, Imp). Should have found a model for girls rather than boys so testing on themselves was an option (Kay, FG).	*Prediction* - to generate a list of rules and statements about their reliability. "We use this like [the dataset]. We figure out a formula to get to that height" (Iona, Imp). "The PSO lacks good height estimation skills. Utilizing their strengths of estimating short lengths, a series of reliable rules of thumb must be evaluated to provide a good height estimate" (Kay Report – similar in other reports).	"Different things (body parts) are proportionate to different. . . . I felt like they weren't all connected" - couldn't just add up randomly (Iona, VSR). "Gender: you would expect all the boys to be on one side and all the girls on the other one. You can use all of them (i.e., variables). That's why they are all there" (Kay, Imp). "Depend on what smaller values they are able to get" (Kay, VSR). "It was kind of not to just have one that would like kind of work for all, to have . . . ways to work out a few of them" (Iona, VSR). "You can see the gender of someone . . . rule can be a bit more specific" (Kay, VSR). Mathematical Modeling aligned to mental model of situation.

Note. Imp – Audio and video recorded lesson implementation, VSR – Video post lesson video stimulated recall interview, FG – focus group interview post lesson.

Table 3. Coding of purposes of use of technology.

Digital Technology Use	Modeling Purposes of Use Code[a]	Modeling Approaches of technology use code[b]	Teams[c]
Home screen calculations on basic function calculator on laptop/CAS calculator	Making choices about features considered significant during pre-mathematization facilitated by digital technology use	N/C	PSO2019 Green(HC); PSO2019 Blue(HC); PSO2019 Orange(HC); PSO2019 Red(HC); PSO2019 Mauve(EC); WLT2019 White(HC); WLT2019 Green(HC)
	Using technology to verify mathematical model output	UUT	PSO2019 Red(HC); PSO2019 Mauve(EC); PSO2019 Purple(EC) use solve to compare predicted with actual height; PSO2019 Pink (EC); **WLT 2019 Red(HC)**; WLT2019 White (HC)
	Using digital technology to Work Mathematically with model to carry out simple calculations such as summation, finding averages and ratios	N/C	PSO2019 Blue (HC); PSO2019 Orange (HC)(+/x only); PSO2019 Red(HC); PSO2019 White(HC); PSO2019 Teal(EC); **PSO2019 Olive(EC)**; WLT2019 Blue(HC); WLT2019 Red(HC) [not Evan]; WLT2019 White(HC)
Using Excel Spreadsheet with Function formulae such as AVG	Using digital technology in building models – constructing the formula (e.g., Ht = SL x AvHt/AvSL) or ratio	SGT	PSO2019 Green(HC); PSO2019 Red(HC); PSO2019 White(HC); PSO2019 Purple(EC); PSO2019 Pink(EC)
Use of scatterplot	To identify error [interchanging variables]	N/C	PSO2019 Pink(EC)
	To identify or get a sense of relationship	SGT	WLT2019 Green(HC); **WLT Red(HC)**
	To confirm expected 'linear' trend [visually] – 'looks pretty much like a line' holding a pencil up to CAS	SGT	WLT2019 Blue(HC).
	Compared graph of model with data points, determine best has 'more dots closer to line'	SGT	WLT2019 Green(HC); WLT2019 Orange(HC)
Use of Scatterplot and Lin Reg oto generate line of best fit	Using digital technology in building models – constructing the formula	SGT	PSO2019 Grey(EC); PSO2019 Mauve (EC); PSO2019 Gold(EC); PSO2019 Silver (EC) [model not recorded]; PSO2019 Purple (EC); PSO2019 Pink(EC); WTL2019 Red(HC)[not Evan]; WLT2019 White(HC); WLT2019 Green(HC); WLT2019 Orange(HC); **PSO2019 Olive(HC)** abandoned for more sense making model grounded in reality
Generation of GoF Values	Recorded/ generated but not used for model selection or comparison	PFG	PSO2019 Grey(EC); PSO2019 Gold(EC) as model with smaller r value (KH 0.71 cf 0.77 for WtE) gave closer predictions; PSO2019 Pink(EC); PSO2019 Teal(EC); **PSO2019 Olive(EC)**
	Used to select the "best" model	FFS	PSO2019 Mauve(EC); PSO2019 Purple(EC)
Sorting data with spreadsheet initially	Sifting through data of a modeling situation facilitated by technology as part of pre-mathematization	N/C	PSO2019 Green(HC)
Sorting data with spreadsheet after first models produced	Using sorted data to refine models produced for the entire dataset for subpopulations of datasets (e.g., based on gender/ethnicity)	SGT	PSO2019 Red(HC)

(Continued)

Table 3. (Continued).

Digital Technology Use	Modeling Purposes of Use Code[a]	Modeling Approaches of technology use code[b]	Teams[c]
Using technological output (e.g., spreadsheet, calculator calculations, graphical displays)	Using technological output to facilitate group work by supporting argumentation between members of group	SGT	PSO2019 Red(HC); **PSO2019 Olive(EC)**; WLT2019 Blue(HC); **WLT2019 Red(HC);** WLT2019 Green(HC)
Tracing along the trend line	Using technological output to facilitate group work by supporting argumentation between members of group – "Just explaining it" (Lyn, VSR). During implementation – traced slope "as your height gets bigger, the knee height . . .the trend. I think while we were talking, we were just tracing it" (Lyn, VSR).	SGT	WTL2019 White(HC); WLT2019 Blue(HC); PSO2019 Mauve (EC)
Using moveable line to find by-eye location of best linear model	Using digital technology in building models - comparing fit of model by eye	SGT	**WLT2019 Red(HC)[Evan]**
Using Excel spreadsheet to store and display manually sorted data	Using digital technology in building models – ratio	UUT	PSO2019 White(HC)
Use of technology enabling use of larger data set	Having access to technology allowed choice of size of sample to investigate (not limited by more time consuming by-hand methods)	N/C	PSO2019 Gold(EC); PSO2019 Silver(EC); **PSO2019 Olive(EC)**
Using internet to find procedural descriptions of how to carry out a technological technique (e.g., sorting, using a moveable line, how to form an equation)		N/C	PSO2019 Green(HC); WLT2019 Green(HC)

Note. AVG Average command; EC Earnshaw College; HC Heathcliff College; FFT Fitted Function Selector; GoF Goodness of Fit; Ht height; N/C no code; SGT Scatterplot/Graphing Tool; SL shoe length; UUT Unneeded/Unused Tool; VSR post lesson Video Stimulated Recall interview.
[a]Initial codes from Villarreal et al. (2018, p. 333); [b]Codes from Zbiek (1998, p, 198); [c]Use of bold indicating teams displaying modeling the phenomenon as approach

Table 4a. Partially Ordered Meta-Matrix: Modeling Processes (Heathcliff College Data).

		Cognitive activities in Transition						
Team	Approach	1	2	3	4	5	6	7
PSO Task								
Green (Una & Tegan)	MD	U, St, S, IntC	IdV, F, M, J	WM, Ver	IntM	Cp, Cq	Com, J (prec.)	RV (same model form but other Var)
Blue (Eric & Ray)	MD	U, St, IntC	F, M	WM	IntM	Cp, Cq, Val	Com, J (prec. – max error)	RV (other models & to refine proportional combination model)
Orange (May, Clare, Ruby)	MD	U, St, IntC	F, M	WM		Cp	Com	
Red (Tom, Pat)	MD	U, St, IntC	F, M	WM	IntM	Cp, Cq	Com, J (consistency)	RV to refine models accounting for ethnicity
White (Ned & Joe)	MD	U, St, IntC, N, S	A, IdV, F, M, J	WM, Ver	IntM		Com	
Brown (Evan & Zald)	MD	U, Int C	F, M	WM, Ver	Int M	Cp, Cq, Val	Com, J (prec.)	RV (same model form but other Var)
Weight Lifting Task								
Green(Una & Tegan)		U, St, IntC	F, M	WM	IntM	Cp, Cq	Com, J (fit)	RV(other models linear&logarithmic)
Blue (Eric & Ray)	MD	U, St, IntC	F, M	WM	IntM	Cp, Cq	Com	
Orange (May, Clare, Ruby)	MD	U	F, M	WM		Cp	Com	
Red (Evan, Tom, Pat)	MD/MP(Evan)	U, St, IntC	F, M	WM	IntM	Cp, Cq	Com	RV (same model form but other Var) Evan does not RV
White (Ned & Joe)	MD	U	F, M	WM	IntM	Cp	Com	

Note. A assuming; Com communicating; Cp comparing: Cq critiquing; F formulating; IdV identifying variables; IntC interpreting context; IntM interpreting mathematical output; J justifying; M mathematizing; MD modelling dataset or part thereof; MP modelling phenomenon; N noticing; RV revisiting; S simplifying; St structuring; U understanding; Val validating; Var Variable(s); Ver verifying; WM working mathematically; prec. precision.

Table 4b. Partially ordered meta-matrix: Modeling processes (Earnshaw College data).

		Cognitive Activities in Transition						
Team	Approach	1	2	3	4	5	6	7
PSO Task								
Gold (Ann, Bria, Carol)	MD	U, St,	F, M	WM	IntM	Cp, Cq	Com	
Silver (Deja, Eve)	MD	U, St					Com	
Purple (Fay, Gen, Honi)	MD	U, St, Int C	F, M	WM	IntM	Cp, Cq	Com	
Pink (Iona, Jen, Kay)	MD	U	A, F, M	WM		Cp	Com	
Mauve (Lyn, Macy, Leita)	MD	U, Int C	F, M (ratio model)	WM	IntM	Cp, Cq	Com	RV (look at refining to Caucasian Female model – reject as inaccurate and low r; test several variables)
Grey (Noni, Ona, Polly)	MD	U, Int C	F, M	WM	IntM	Cp, Cq	Com, J (prec.)	RV (reject an incorrect model, same model - using other variables)
Teal (Quin, Rosa)	MD	U, Int C	A	WM		Cp, Cq	Com	
Olive (Sui, Tara, Veta)	MP	U, St, S,	F, M	WM	IntM	Cp, D	Com	RV (other models)

Note. A assuming; Com communicating; Cp comparing; Cq critiquing; D deciding; F formulating; IntC interpreting context; IntM interpreting mathematical output; J justifying; M mathematizing; MD modeling dataset or part thereof; MP modeling phenomenon; RV revisiting; S simplifying; St structuring; U understanding; WM working mathematically; prec. precision.

Ned: I feel like I kind of narrowed myself down, just by viewing it as the whole goal was to find these two [world standards needed], because then I wasn't really having a model of humans doing weight lifting, I was just having a model like based on the averages of numbers to estimate this number [points to world standard needed], like yeah. And that was my mistake.

Initially, he had expected that there would be a "trend between how much they weigh and how much they can lift." To find this, he suggested to his partner, Joe, using ratios "so it is like for each kg they weigh more than 55, . . . the weight they can lift increases by a set amount." By omitting the last given data point for the superheavy weight lifter, he was able to say that for every 6 kg increase in weight, the lifters lift an extra 9.25 kg. They decided to use Excel on a laptop to make a scatterplot and find a linear trendline. Ned then realized he could use his weight lifted to extra weight of the lifter ratio as the gradient for his own linear model, but could not find the constant by hand techniques, so finally used Joe's constant from Excel at the teacher's suggestion. To complete the task with a report, Ned decided to continue using ratio calculations on his calculator and Joe would use Excel to produce a series of linear models. From this point on, their "modeling" was driven by mathematical expediency, as Ned and Joe explained in the post task interview.

Ned: Obviously there is a limit to what humans can hold, because . . . if any human can carry any weight, then it wouldn't need robots . . . But we did understand, and like looking at the thing it was, there probably was a limit, and we probably should have taken that into account. However, it was the quickest way to have a linear.

Joe: I was just thinking like, instead [put] those parts into the Excel thing and then formulate like the trend line, and find the equation and stuff. . . .. I guess we didn't really like put in the consideration for like the human limit that much, I think.

Interviewer: So you really chose it on mathematical grounds and convenience and efficiency at the time?

Joe: Yeah, pretty much.

Interviewer: So the actual real world didn't come a lot into your thinking?

Joe: I guess not.

Interviewer: You weren't really thinking a lot about weight lifting during it?

Ned: Nuh [meaning "no"].

In a second example, when again the dataset was perceived to be a set of numerical values where contextual meaning was secondary, tensions between flawed physical theories of what was occurring in the situation, the modeler's mental image of it, and the results from mathematical working using actual values, were overridden. In the *WLT*, Eric and Ray (Blue team, Heathcliff College) both decided quickly that they should put the data values into a graph and obtain a trend line, as that would mean "you can say roughly what it is going to be" (Ray). Eric used a CAS calculator to plot the men's World Record Lifts against weight categories, which they decided looked reasonably linear, except for the last point, which was the second highest data pair given, as he left out the super-heavyweight category. Ray drew a line on the plot of the women's World Record Lifts against weight categories on a spreadsheet screen image saying "this one looks pretty much right on the line, yeah." For this line, he used a point consisting of the mean of the weight categories and the mean of the weight lifted (62.3, 140.3), with the point representing the lowest weight category and its corresponding World Record Lift (49, 120), to find the gradient and then the linear equation using $421/3 = 17/11 \times 187/3 + c$. Their model to find missing World Standards for Women was $y = 17/11x + 44$ where y = weight lifted and x = weight class. They then decided to use linear models with the superheavy category missing as their first models, and the actual weights with all data as a second model, not realizing this was just a different data source, not necessarily giving rise to a different mathematical model.

When Eric plotted the men's actual weights with the amount lifted for World records on his CAS calculator, it was not linear (see Figure 2a) with Ray commenting he didn't like it. He noted that, "This one almost looks like it is going to be like a parabola or something." Eric saw it as "going to come back around," but Ray insisted "it is not going to start coming down," suggesting "it is like exponential" and so "turns really gradually," but apart from the last two points, it was "a pretty straight line."

After being reminded by their teacher of how they could find the equation and look at different graph types to use as models on an Excel spreadsheet, they examined the female data on a spreadsheet and various trend line choices, noting that the data plot "is not linear at all" (Ray), but still chose none. Ray proffered the flawed physical theory "if the weight kept going up, like in terms of muscle mass, they will be able to keep lifting more and more as well." For this reason, they were happy with the linear model as their first choice and their discarding of the super-heavyweight class. Eric even suggested it could be "a symbol of underperformance," that the actual data was so much lower than a linear model would predict. Although Ray was willing to consider an exponential model, which he then rejected, he insisted a parabola "is definitely not right at all," despite the better visual fit to the points (see Figure 2b).

As the following exchange shows, there was a mismatch between the boys' mental model of the situation, the physical theories they held about the physiology of the situation, and the mathematical object (the parabola) that appeared to be the best model:

Eric: I mean it doesn't fit in with like how reasonable it would be.

Ray: If you keep on getting like, like if this [the parabola] just goes on, then you get like heavier, you get more muscle, you can lift less until it is nothing!

Eric: That would mean there is a point at which it peaks.

Ray: Yes, but I doubt it. It just doesn't seem right. [softly] I don't like it, it is bad.

A further exchange with the teacher convinced Eric there might come a point where a person had too much muscle, or was too big, and would have difficulty standing up, let alone lifting weights. Reluctantly, Ray accepted the quadratic model, but said they were only choosing it because it "seems to be more accurate." Even in the follow-up interview, Ray said he was not expecting a parabola because "like the way it is kind of formatted, it is like you get stronger, as you like gain weight then you get stronger; but then you keep getting it [more weight] until you get weaker again. And that just didn't make a lot of sense to me, like in a literal sense."

This team's actions do not indicate they were a slave to the data – more there was a misalignment between the situation, Eric's mental model of the situation based on his physical theories about human capacity to lift weight in relation to bodyweight, and the mathematical objects at his disposal to model the situation. All their actions and comments were consistent with modeling the dataset given, although they were willing to exclude real data (i.e., the super-heavyweight category) that did not fit with the physical theories they held, especially Ray's.

Modeling part of dataset not the phenomenon – believing model more general

In the *PSO Task*, Una and Tegan (Green team, Heathcliff College) constructed the model, $6S = H$ (where S is shoe length, H is a person's height, both in mm), as a rule of thumb to determine height from a small subset ($n = 7$) of the given dataset ($n = 41$) from the youngest (4 people), middle (2 people) and oldest age groups (1 person), using ratios of average height ÷ average shoe length. This meant their sample was mainly Caucasian, although Una inspected the data later and declared that there was little variation, so they would not need to adjust their model, even if they restricted it to female Caucasians. They indicated in the follow-up interview that the model applied generally to the adult population "because they are people" (Tegan). Una qualified this by saying, "Maybe not for children because children are a bit different. The growth rate is much different from ours ... from adults." Shoe length had been chosen initially: "because shoe length ... it's better to the person's overall height, like the bigger shoe is obviously taller, because you need bigger shoes to help balance the person more as they weigh more" (Una). They sifted through the given data, making choices about which variables they considered significant, based on their physical theories about growth, eliminating face height as irrelevant, but investigating knee height, as it contributed to a person's growth and therefore height. Using ratios of averages again, they constructed the model $H = 3\,k$ (where k is knee height, H is a person's height, both in mm). From a trial of 5 people's heights from the dataset, they concluded that use of shoe length was the better estimate.

Modeling the phenomenon – believing model more general

In this case, the modeler believed that the model being constructed applied generally. In the *WLT*, for example, Evan from Red team at Heathcliff College believed that the model being constructed applied to the phenomenon of human capacity to lift weight depending on bodyweight, rather than the phenomenon of weight lifters' capacity to lift weight relative to their bodyweight in Olympic competition. He therefore neglected the effects of human decision-making on qualifying benchmarks to be able to participate in the competition, ensuring that the minimum bodyweight was not zero. Evan attempted several times to find what he called the "best matching line." He used a moveable line on his CAS calculator screen showing a joined plot of the world record lifts for males against their estimated or real weight (Figure 3a) with the line locked to pass through (0, 0) (Figure 3b). He anticipated, before he plotted the points, that these would be roughly linear starting from zero, "because weight sort of starts at zero." He realized that the model he produced was only "accurate" for the lower weight categories, as he had removed the others; because, he said, if he included them then the lower categories did not fit his linear model so well.

He acknowledged a linear model did not fit all the data, "because it curves, and there will be a limit on it [running the tips of the fingers of his closed hand across the points in a curved fashion, Figure 3c], like how much it will weigh. And like it is not going to keep a constant rise like this [running his fingertips up the first 5 points, Figure 3d], because you can't like, you can't lift it."

Modeling the phenomenon–believing model applies generally

Finally, there was only one team from either site who were modeling the phenomenon and believed that was what they were doing. Olive team (Sui, Tara, Veta) from Earnshaw College was looking for a rule that covered everyone for the *PSOs*, not just the dataset. They based their modeling on their own

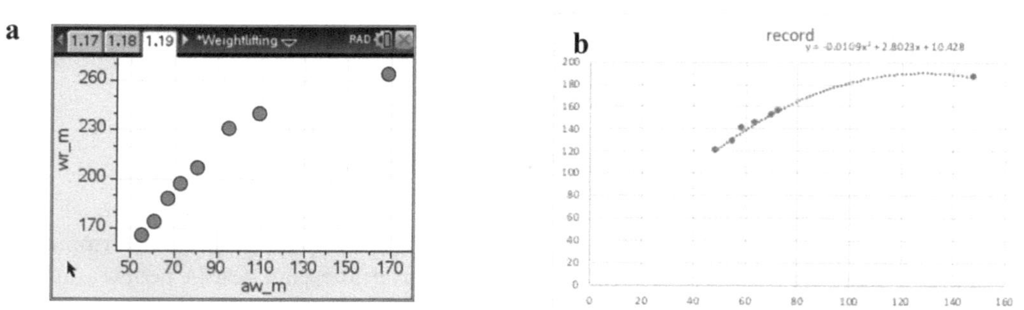

Figure 2. a Eric's plot of weight lifted vs actual weight of male record holders. **b** Ray's quadratic model for weight lifted vs actual weight of females.

measurements and a small subset of the dataset. Following an initial exploration of any physical theories they knew about body part lengths, such as "your shoe length is the same as your forearm" (Tara), they started with linear regression of the full dataset, as this strategy was elicited by the teacher in a well-orchestrated whole class discussion and she had later dismissed a question from Tara about how to show an error interval in a measurement as being off track. They generated a scatterplot on Tara's CAS calculator, but did not understand it, as she had plotted knee height versus knee height. So, they returned to looking at their body proportions and the relationship of these to their height. Tara indicated they planned to look at knee height (Kh) and then wrist to elbow (WtE), to estimate overall height (H).

Tara had shown her team her wingspan during the initial whole class discussion as an example of a rule of thumb saying, "That's how tall you are." They now looked at the length they needed as consisting of their forearms, which should be the same lengths, similarly with upper arm lengths and then their torso, which Sui pointed out was unknown. They started making estimates such as Arm = 2WtE and suggested checking by measuring on themselves. Sui suggested doubling the WtE could find the demi-span, which they could also test by measuring, as demi-spans were not in the given dataset.

Tara returned to using the scatterplot and linear regression on the CAS calculator after a visit to their team by the teacher. A comment by the teacher about recording what was in each list triggered, for her, the reason that their attempt at linear regression was unsuccessful. She generated and recorded the linear model and its goodness-of-fit values and her interpretation of these as: $y = -170.5 + 0.4x$, where y = height and x = knee height, $r^2 = 0.67$ = moderate. For Tara, moderate for "the strength of the data" meant the data "isn't very accurate" but, more telling, was the fact that the team still did not understand what it showed them, as it "didn't make any sense" (Tara, VSR).

From this point on, they "used our own bodies and figured it out from there" (Tara, VSR). Arm = 2WtE led to Demi-span = 3WtE to predict height. Testing on Tara and Veta gave mixed results, but an exchange between Veta and Tara indicated Veta expected that any models they made could not exclude short people like herself, so their idea of stating the degree of error expected as an interval, was reiterated. They then returned to considering knee height with Sui observing, "if you times your knee height by four it kind of gets your height." For Alice, the first person in the dataset, Sui had found the ratio, H/Kh = 3.7. They started to test this on the given dataset, but moved on to wrist to elbow, as the teacher told the class they should be onto their second model. Both 5 and 4 were suggested as multipliers for knee height to estimate height, but Tara wanted what they were doing grounded in reality, so brought them back to relating what they were doing to their wingspans. Sui suggested that Tara's chest was about the same length as her forearm. They then decided to get an idea of the width of torsos by finding 4 x WtE, which corresponded to the length of the arms, and compared this to the heights of Asha, Bindi, and Alice in the dataset, finding differences of 540, 560, and 550 mm, respectively. The model for height was then $H = 4WtE + c$, where c was the average of widths of torso (i.e. the average difference between the three predicted total length of arms and actual heights in their comparisons). They tested on themselves empirically by measuring, believing the model should work generally. Finally, two rules relating wrist to elbow and height were proposed, one for males and one for females: $H_M = 4WtE + 750$ and $H_F = 4WtE + 550$, as they expected males to have wider torsos, so allowed an extra 200 mm. The rule for females they again tested on themselves empirically. They noted the need to expect variation and so "give or take 100 millimetres" when using the models. In this instance, the modeling was driven mainly by a desire to engage with the underlying phenomenon by understanding the relationships existing between the variables in the situation, informed by wanting to integrate their personal theories about the physiology of the situation with the mathematical objects they used to model it, in a way that made sense to them.

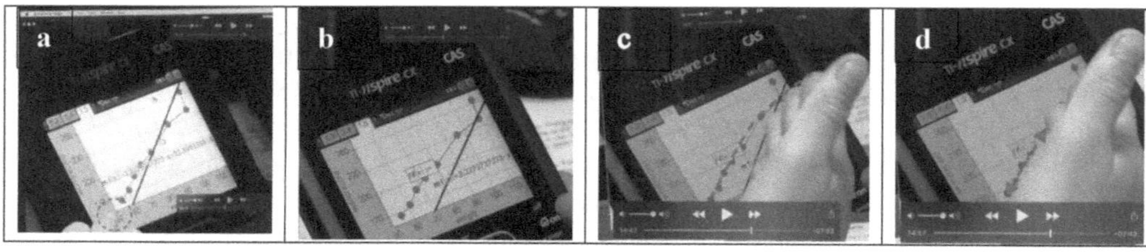

Figure 3. **a** & **b** Evan fitting a "best matching line" passing through (0, 0). (Note: Last data point is incorrectly entered.) **c** & **d** acknowledging top weight categories do not fit his linear model produced with the lower weight categories.

Influence of digital tools on data modeling approaches to the modeling tasks

To gauge whether there was any influence of digital tools, with curve fitting and regression capabilities, on students' choice of data modeling approach, we turn our attention to Table 3. The two teams, where members were engaged in modeling the phenomenon, both used the CAS calculator. Evan from Red team started immediately by plotting a scatterplot of the weight categories of male weightlifters and their world record lifts on his CAS, expecting that he could use the fitting of a line of best fit by eye by using the moveable line facility to construct a linear model in the *WLT*, and so predict the missing world standards. The availability of the calculator to do the plot enabled construction of an initial model and further model exploration. However, the dataset was small and the scatterplot, fitting of the line, and finding an equation for it could have been done by hand or missing values read from the line. Olive team generated a linear model in their attempt at solving the *PSO* task using linear regression with goodness-of-fit values; but abandoned this approach for one where they were able to make sense of their modeling using their understandings of the relationships between variables in the context, based on proportional models.

With respect to Zbiek's (1998) modeling approaches of technology use, there were just two examples of use of the Fitted Function Selector approach, where final model choice was selected on the basis of Goodness-of-Fit values, rather than any knowledge of the real situation. Both teams used a modeling the dataset approachand treated the task as practice in function-fitting rather than as an opportunity to build and validate a model. Five teams used the Potential Function Generator approach for at least part of their modeling with some aspects of the real world being used to select or reject their model(s). Given that the purpose of their modeling was seen by virtually all groups as prediction, the high incidence of Scatterplot/Graphing Tool approach is unsurprising. This was used by teams using either data modeling approach. With respect to RQ1b, having access to technological tools allowed either approach to data-modeling to be enacted and enabled the use of least squares regression, which no student was able to do by-hand, as a by-hand approach was not part of their learning experiences. The tools allowed several models to be considered in graphical form easily within the timeframe allocated, a point several students made in the focus group and VSR interviews.

Modeling processes engaged in when modeling these data-rich situations

Tables 4a and 4b present the approach taken (i.e., modeling the dataset or modeling the phenomenon), and the modeling processes engaged in by students at the two sites whilst modeling these data-rich situations as a partial answer to RQ2. Each set of data for a team was coded for instances of cognitive activities in the transitions in the modeling process shown in Figure 1. Some additional cognitive activities related to modeling processes were also coded (see Table notes). Further details of cognitive activities in these transitions are evident in the modeling purposes of use column of Table 3.

For transition 1, there were only two incidences of simplifying, one in a team fully aware of using the modeling the dataset approach and one by a team fully aware they were using a modeling the phenomenon approach. Both occurred in the *PSO Task*. In transition 2, the making of assumptions was very low and there were no incidences of making assumptions in any other transition. The three occurrences were in the *PSO task* with teams taking a modeling the dataset approach. There remained an imperative for teams to understand relationships between relevant variables in the data with either approach. Structuring of the messy real-world situation was not always evident but when it was, it was based on recalling experiential knowledge. Structuring was identified in modeling of both tasks with teams employing either approach. Working mathematically was evident for all teams, with some verifying output of their models from a mathematical perspective as well, enabled for several teams by access to digital technology (see Table 3). Interpreting mathematical output was evident in 14 of the 19 team attempts at the tasks. In transition 5, comparing and, to a slightly lesser extent, critiquing of different models occurred in most teams but validating, in the sense of reflecting on their models in light of the problem posed in the real situation, and evaluating that the correct model had been constructed for it, was rare.

In transition 6, several teams rejected reasonable models in terms of fit and realistic match to the situation based on being too inaccurate when extrapolating beyond the data (for larger values of the independent variable), whilst simultaneously ignoring similar issues when extrapolating for smaller values. This occurred in both tasks and for both data-modeling approaches. The last transition highlighted difficulty with revisiting the model as several teams did not do so, as their time management did not allow, they did not know how to, or they were not inclined to want to engage further in doing any more mathematics. However, of the teams who did attempt to revisit, several merely revisited the same model using different variables, not a new model, or for the refinement of an existing model. Again, this occurred in both tasks and with both approaches.

Discussion and conclusion

The aim of this paper was to characterize how students working in small teams act when working with data-rich modeling tasks in order to develop meaning for their modeling and the models they produce. In particular, the differences between modeling the dataset and modeling the phenomenon were explored to add to the sparse reporting and discussion of this in the literature to date (e.g., Galbraith, 2007; Riede, 2003). Riede's examples were all in sports contexts and were the inspiration for the *Weight Lifting Task*. However, the *PSO Task* is from a non-sports context, as was Galbraith's daylight example. Most student teams in this study took a modeling the dataset approach, but the degree they were aware of this differed from a deliberate choice to do so, and acknowledgment of this, to believing they were not doing this but instead, modeling the phenomenon itself.

As the students were learning to model and to develop meta-knowledge about mathematical modeling in technology-active classrooms, it was pleasing to see many teams looking for relationships within the situation guided, to some extent, by their personal physical/physiological theories about the interplay of what they considered relevant and irrelevant variables as proposed previously by Czocher and Hardison (2021). When there was more alignment between these personal theories, espoused mental models (Waisel et al., 2008) of the situation (although only a few incidences of these were revealed in the collected data) and the mathematical modeling being employed, the quality of the modeling was higher. The presence of technology-generated external visualizations such as scatter-plots, or sorted displays of categorized derived values such as ratios and averages, allowed students to discuss these mental models as they externalized their thinking and augmented argumentation within their teams about suitability for representing the situation.

With respect to the modeling processes engaged in by students whilst modeling these data-rich situations, the foreshadowing by Caron in her polemic chapter (Caron, 2018) that use of such tasks would lead to a quite limited experience of the processes of modeling and rely on little development of an understanding of the real situation being modeled was not borne out for most teams as is shown in Tables 4a and 4b and exemplified in the findings section. For some transitions, incidence of particular cognitive activities, namely, simplifying, making assumptions, verifying, and validating was less than ideal, nevertheless. Lack of particular processes was not unequivocally associated with either approach. However, it might be that the tasks used did not evoke enough examples of modeling the phenomenon for any trends to be seen, so further research in this area is warranted.

With respect to *theory development*, we have been able to identify incidences of both types of approaches to data-rich situations in these modeling tasks. The relatively low incidence of modeling the phenomenon identified here may, however, be an artifact of the data used in the study, or its local context, rather than a real one. Data-rich modeling can, and should, involve considered use of data but also accounting for knowledge of the problem context (Galbraith, 2007; Riede, 2003).

From an *educational practice perspective*, data-rich modeling tasks as used here, unlike other forms of modeling tasks, allow modelers to focus on a dataset as an instance of a phenomenon and/or as the phenomenon itself informed by scrutiny and analysis of this particular instance. Only focusing on the dataset limits concentration on the phenomenon, but it also provides a range of options for initiating modeling. It thus allows informed class discussions to contribute to students' meta-knowledge about mathematical modeling by considering whether students are modeling the dataset or modeling the phenomenon, their awareness of the data-modeling approach taken and their thinking about the generality and area of application of the models under construction. There is also an opportunity to discuss as a whole class, or within group facilitation, how a model of the data might not easily translate to an explanation of why the model aligns with the situation being modeled.

Students have access to multiple functions in technology-active environments, often before they know much about the various function types typically in secondary education. Teachers can, and should, provide opportunities for students to make sense of how technological functions of tools such as CAS calculators and spreadsheets with in-built trendline choices can be used sensibly when modeling. With data-rich tasks, it seems eminently sensible to produce plots of some, or all, of the data and reflect on these (Brown, 2015). Depending on one's knowledge of the context under investigation, these plots can confirm anticipated relationships, or suggest relationships, that should then be considered in terms of the context. Given data-rich tasks have reasonable sized datasets, using tools such as regression with some thought to function selection can empower users.

An increased emphasis by teachers on students better communicating their reasoning and decision-making as they solve real-world tasks is warranted. A tendency to communicate once finished modeling, or time was running short, restricted opportunities for students to recognize errors, resolve different understandings in the team, and emphasize connection between the mathematical and real worlds (Pollak, 1997). This emphasis may also result in students becoming more aware of their data-modeling approach as part of the meta-knowledge about mathematical modeling students are expected to develop when using data-rich modeling tasks of this nature. It may help them keep focus on the real-world problem and not become stuck in the mathematical world, forgetting the purpose of solving the task.

It is important that students reflect on their modeling and model(s) produced (Maull & Berry, 2001; Yoon & Thompson, 2007) and engage in the cognitive activities associated with the transitions from accepting of the solution and reporting (transition 6) and revising the model and returning to the real-world situation (transition 7). To this end, curricula documents should include increased emphasis on the domain over which a function might be useful. The tendency to reject reasonable models, both in terms of fit and realistically matching the real-world situation, might then be counteracted. However, the Fitted Function Selector approach (Zbiek, 1998) of seemingly blindly accepting a regression model and also selecting best on the basis solely of a statistical measure must be discouraged as not appropriate when modeling. Students and teachers need to be more aware that the closeness of fit of selected data to a function does not necessarily reflect reality. Taking a regression model as a starting point can be appropriate where needed, using functionality such as moveable lines (as used by Evan in Red team), but also recognizing that CAS-enabled technologies generally allow users to also dynamically manipulate a graphical model to reflect real-world considerations and represent plotted data. This suggests that with today's digital technology, a merging of the Potential Function Generator and ScatterPlot/Graphing Tool (Zbiek, 1998) should be the approaches encouraged as they are in keeping with expanding students' meta-knowledge about modeling. The results from this study depend on the teaching and learning context, the actual tasks used, and the work of the teams selectively reported. This must be borne in mind in interpreting them.

Notes

1. This research was supported by the Australian Government through Australian Research Council's *Discovery Projects* funding scheme (DP17010555). The views expressed herein are those of the authors, not necessarily those of the Australian Government or Australian Research Council. Chief Investigators are V. Geiger, G. Stillman, J. Brown, and P. Galbraith. M. Niss is a Partner Investigator.
2. All names used for schools and students are pseudonyms.
3. Fuller versions available at www.mathsmodellingenablers.com

Disclosure statement

No potential conflict of interest was reported by the author(s).

Funding

This work was supported by the Australian Research Council [DP17010555].

ORCID

Gloria Ann Stillman http://orcid.org/0000-0002-0996-9120
Jill P. Brown http://orcid.org/0000-0002-9202-3634

References

Association of Mathematics Teacher Education. (2017). *Standards for preparing teachers of mathematics*. Available from http://amte.net/standards

Barquero, B., & Jessen, B. E. (2020). Impact of the theoretical perspectives on the design of mathematical modelling tasks. *AIEM - Avances de Investigación En Educación Matemática, 17*, 98–113. https://doi.org/10.35763/aiem.v0i17.317

Bautista, A., Wilkerson-Jerde, M. H., Tobin, R. G., & Brizuela, B. M. (2014). Mathematics teachers' ideas about mathematical models: A diverse landscape. *PNA, 9*(1), 1–28 http://hdl.handle.net/10481/33231.

Berry, J. (2002). Developing mathematical modelling skills: The role of CAS. *ZDM-Mathematics Education, 34*(5), 212–220. https://doi.org/10.1007/BF02655824

Blum, W., & Leiß, D. (2006). "Filling up" – The problem of independence-preserving teacher interventions in lessons with demanding modelling tasks. In M. Bosch (Ed.), *Proceedings of the fourth conference of the european society for research in mathematics education*. Guixol.

Borromeo Ferri, R. (2018). *Learning how to teach mathematical modeling in school and teacher education*. Springer.

Brown, J. P. (2015). Visualisation tactics for solving real world tasks. In G. A. Stillman, W. Blum, & M. S. Biembengut (Eds.), *Mathematical modelling in education research and practice: Cultural, social and cognitive influences* (pp. 431–442). Springer.

Caron, F. (2018). Modelling in secondary mathematics education: Moving beyond curve fitting exercises. In A. Kajander, J. Holm, & E. Chernoff (Eds.), *Teaching and learning secondary school mathematics: Canadian perspectives in an international context* (pp. 547–560). Springer.

Czocher, J. A., & Hardison, H. L. (2021). Attending to quantities through the modelling space. In F. K. S. Leung, G. A. Stillman, G. Kaiser, & K. L. Wong (Eds.), *Mathematical modelling education in East and West* (pp. 263–272). Springer.

Docter, P. (2009). *Up [Film]*. Pixar Animation Studios.

Doerr, H. M., Ärlebäck, J. B., & Misfeldt, M. (2017). Representations of modelling in mathematics education. In G. A. Stillman, W. Blum, & G. Kaiser (Eds.), *Mathematical modelling and applications: Crossing and researching boundaries in mathematics education* (pp. 71–81). Springer.

Gainsburg, J., Fox, J., & Solan, L. M. (2016). Argumentation and decision making in professional practice. *Theory into Practice, 55*(4), 332–341. https://doi.org/10.1080/00405841.2016.1208072

Gainsburg, J. (2013). Learning to model in engineering. *Mathematical Thinking and Learning, 15*(4), 259–290. https://doi.org/10.1080/10986065.2013.830947

Galbraith, P. (2007). Dreaming a 'possible dream': More windmills to conquer. In C. Haines, P. Galbraith, W. Blum, & S. Khan (Eds.), *Mathematical modelling (ICTMA 12): Education, engineering and economics* (pp. 40–62). Horwood.

Geiger, V., Stillman, G., Brown, J. P., Galbraith, P., & Niss, M. (2018). Using mathematics to solve real world problems: The role of enablers. *Mathematics Education Research Journal, 30*(1), 7–19. https://doi.org/10.1007/s13394-017-0217-3

Gilbert, A., & Suh, J. M. (2021). Convergent nature of modelling principles across the STEM fields: A case study of preservice teacher engagement. In J. M. Suh, M. H. Wickstrom, & L. D. English (Eds.), *Exploring mathematical modelling with young learners* (pp. 345–372). Springer.

Greefrath, G. (2020). Mathematical modelling competence: Selected current research developments. *Avances de Investigación En Educación Matemática, 17*, 38–51 https://doi.org/10.35763/aiem.v0i17.

Jensen, J. H., Niss, M., & Jankvist, U. T. (2017). Problem solving in the borderland between mathematics and physics. *International Journal of Mathematical Education in Science and Technology, 48*(1), 1–15. https://doi.org/10.1080/0020739X.2016.1206979

Krüger, A., Vorhölter, K., & Kaiser, G. (2020). Metacognitive strategies in group work in mathematical modelling activities—The students' perspective. In G. A. Stillman, G. Kaiser, & C. E. Lampen (Eds.), *Mathematical modelling education and sense-making* (pp. 311–321). Springer.

Maull, W., & Berry, J. S. (2001). An investigation of student working styles in a mathematical modelling activity. *Teaching Mathematics and Its Applications, 20*(2), 78–88. https://doi.org/10.1093/teamat/20.2.78

Miles, M. B., Huberman, A. M., & Saldaña, J. (2014). *Qualitative data analysis: A methods sourcebook* (3rd ed.). Sage.

Miles, M. B., & Huberman, A. M. (1994). *Qualitative data analysis: A methods sourcebook* (2nd ed.). Sage.

National Governors Association Center for Best Practices and the Council of Chief State School Officers (2010). *Common core state standards for mathematics*. http://www.corestandards.org/assets/CCSSI_Math%20Standards.pdf

Niss, M., Blum, W., & Galbraith, P. (2007). Introduction. In W. Blum, P. L. Galbraith, & H.-W. Henn (Eds.), *Modelling and applications in mathematics education: The 14th ICMI study* (pp. 3–32). Springer.

Pollak, H. (1997). Solving problems in the real world. In L. A. Steen (Ed.), *Why numbers count: Quantitative literacy for tomorrow's America* (pp. 91–105). College Entrance Examination Board.

Richards, L. (2005). *Handling qualitative data: A practical guide*. Sage.

Riede, A. (2003). Two modelling topics in teacher education and training. In Q.-X. Ye, W. Blum, S. K. Houston, & Q.-Y. Jiang (Eds.), *Mathematical modelling in education and culture* (pp. 209–222). Horwood.

Smith, C., & Morgan, C. (2016). Curricular orientations to real-world contexts in mathematics. *The Curriculum Journal, 27*(1), 24–45. https://doi.org/10.1080/09585176.2016.1139498

Stillman, G. A., Brown, J. P., & Czocher, J. (2020). Yes, mathematicians do x so students do x, but it's not the x you think. *ZDM - Mathematics Education, 52*(6), 1211–1222. https://doi.org/10.1007/s11858-020-01183-5

Stillman, G. (2011). Applying metacognitive knowledge and strategies in applications and modelling tasks at secondary school. In G. Kaiser, W. Blum, R. Borromeo Ferri, & G. Stillman (Eds.), *Trends in teaching and learning of mathematical modelling* (pp. 165–180). Springer.

Stohlmann, M., DeVaul, L., Allen, C., Adkins, A., Ito, A., Lockett, D., & Wong, N. (2016). What is known about secondary grades mathematical modelling—A review. *Journal of Mathematics Research*, *8*(5), 12–28. https://doi.org/10.5539/jmr.v8n5p12

Thompson, M., & Yoon, C. (2007). Why build a mathematical model? Taxonomy of situations that create the need for a model to be developed. In R. A. Lesh, E. Hamilton, & J. J. Kaput (Eds.), *Foundations for the future in mathematics education* (pp. 193–200). Routledge.

Victorian Curriculum and Assessment Authority. (2015). *Victorian certificate of education mathematics study design (2016-2020).* VCAA.

Victorian Curriculum and Assessment Authority. (2016). *Victorian curriculum mathematics: Foundation–10.* VCAA.

Villarreal, M. E., Esteley, C. B., & Smith, S. (2018). Pre-service teachers' experience within modelling scenarios enriched by digital technologies. *ZDM-Mathematics Education*, *50*(1–2), 327–341. https://doi.org/10.1007/s11858-018-0925-5

Vorhölter, K., Kaiser, G., & Borromeo Ferri, R. (2014). Modelling in mathematics classroom instruction: An innovative approach for transforming mathematics education. In Y. Li, E. A. Silver, & S. Li (Eds.), *Transforming mathematics instruction: Multiple approaches and practices* (pp. 21–36). Springer.

Vorhölter, K. (2018). Conceptualization and measuring of metacognitive modelling competencies: Empirical verification of theoretical assumptions. *ZDM-Mathematics Education*, *50*(1–2), 343–354. https://doi.org/10.1007/s11858-017-0909-x

Vos, P. (2015). Authenticity in extra-curricular mathematics activities: Researching authenticity as a social construct. In G. A. Stillman, W. Blum, & M. S. Biembengut (Eds.), *Mathematical modelling in education research and practice: Cultural, social and cognitive influences* (pp. 105–113). Springer.

Waisel, L. B., Wallace, A. A., & Willemain, T. R. (2008). Visualisation and model formulation: An analysis of the sketches of expert modelers. *Journal of the Operational Research Society*, *59*(3), 353–361. https://doi.org/10.1057/palgrave.jors.2602331

Yoon, C., & Thompson, M. (2007). Cultivating modelling abilities. In R. A. Lesh, E. Hamilton, & J. J. Kaput (Eds.), *Foundations for the future in mathematics education* (pp. 201–209). Routledge.

Zbiek, R. M., & Conner, A. (2006). Beyond motivation: Exploring mathematical modelling as a context for deepening students' understandings of curricular mathematics. *Educational Studies in Mathematics*, *63* (1) , 89–112. https://doi.org/10.1007/s10649-005-9002-4

Zbiek, R. M. (1998). Prospective teachers' use of computing tools to develop and validate functions as mathematical models. *Journal for Research in Mathematics Education*, *29*(2), 184–201. https://doi.org/10.2307/749898

Does Strategic Knowledge Matter? Effects of Strategic Knowledge about Drawing on Students' Modelling Competencies in the Domain of Geometry

Johanna Schoenherr (née Rellensmann), Stanislaw Schukajlow, Judith Blomberg and Claudia Leopold

ABSTRACT

We investigated how students' strategic knowledge about situational (pictorial) drawings and mathematical (schematic) drawings affects drawing accuracy and modeling competencies in the domain of geometry. We conducted a pre-posttest experimental study with 473 students in grade 9. Students were randomly assigned to one of three treatment conditions in which we used a 90-minute intervention to promote strategic knowledge about situational drawing (EG1), strategic knowledge about mathematical drawing (EG2), or strategic knowledge about situational and mathematical drawing (EG3). We also used a control group in which we did not promote any knowledge about drawing (CG). Results of a multilevel path analysis did not show a total effect of the strategic knowledge treatments on students' modeling competencies. However, the results indicated an indirect effect: Students who participated in the treatments demonstrated higher modeling competencies than students in the control condition, and strategic knowledge and drawing accuracy were mediating variables. Moreover, students who constructed a more accurate situational or mathematical drawing for a modeling problem were more likely to solve this problem adequately. Our findings indicate that strategic knowledge is a necessary but not sufficient precondition for the construction of high-quality drawings and high modeling competencies in geometry.

Introduction

Initiating mathematical modeling activities in mathematics instruction is aimed at developing students' competencies in using mathematics in the real world. Such competencies are considered important for students' participation in work and everyday life (English & Sriraman, 2010). In recent years, research has been conducted on how to help students develop their modeling competencies, and researchers have called for more intervention studies on modeling (Stillman et al., 2017). One approach is to trigger the use of self-generated drawings by instructing students to make a drawing. Although self-generated drawing is considered a powerful tool for modeling in the domain of geometry (Hershkowitz et al., 1989), students do not generally benefit from instructions to make drawings for the topics of proportional and non-proportional reasoning (De Bock et al., 1998) and modeling (Bräuer et al., 2021). One explanation is that students' drawings are not very accurate and are thus not helpful for solving the modeling problems (Bräuer et al., 2021). One important precondition for drawing accuracy (i.e., the degree of correspondence between the modeling problem and its representation in the drawing) might be strategic knowledge about drawing (i.e., knowledge about the characteristics of a drawing that would fit the modeling problem) (Rellensmann et al., 2020). We report an experimental study in which we investigated how strategic knowledge about different types of drawings affects the accuracy of students' drawings and their modeling competencies in the domain of geometry.

Theoretical background

Mathematical modeling

The term *mathematical modeling* refers to the use of mathematics to understand, structure, and solve problems in the real world (Niss et al., 2007). Mathematical modeling is considered a way to promote the understanding of mathematics and to enable students to use mathematics in everyday life (English & Sriraman, 2010). Modeling activities can be initiated by using modeling problems. An exemplary modeling problem from the domain of geometry is given in Figure 1.

In accordance with classifications of modeling perspectives (Kaiser & Sriraman, 2006), we adopted a cognitive perspective on modeling aimed at understanding the cognitive processes and conditions of the effective use of drawings to solve geometry modeling problems. In modeling research, theories specific to modeling have evolved, including the theory of modeling competencies (Geiger & Frejd, 2015). Modeling competencies include students' abilities, skills, and willingness to work autonomously and insightfully through the solution of a modeling problem (Blomhøj & Jensen, 2003; Maaß, 2006). Whereas (modeling) performance refers to a student's solution to a specific (modeling) problem, a student's average (modeling) performance in solving a sample of different modeling problems is used as an indicator of their (modeling) competencies (Klieme et al., 2008).

Commonly, the process of solving a modeling problem can be summarized with a schematized and idealized cycle that includes key activities (e.g., Blum & Leiss, 2007; Galbraith & Stillman, 2006): In the first step, the student has to *understand, simplify, structure*, and *idealize* the problem to build a model of the situation. For example, the student needs to read the fire-brigade modeling problem (Figure 1), take into account the fire engine, turn-ladder, and rescue object, and make an assumption about the engine's parking position, which could be that the engine is parallel-parked on the street. Also, the student has to select the relevant information, which could consist of the distance needed for safety, the engine's width and height, the length of the ladder, and the rescue height (Figure 2). In the second step, *mathematizing* leads to a mathematical model. For example, the student can mathematize the searched information as the side of a right-angled triangle and an attached distance. He or she can then apply the Pythagorean theorem to formulate an equation. Next, *working mathematically* within the mathematical model produces a mathematical result. For example, the student can solve the equation through a transformation or can measure the required distance in a true-to-scale drawing and multiply

Fire brigade

In 2004, the Munich fire brigade got a new fire engine with a turn-ladder. Using the cage at the end of the ladder, the fire brigade can rescue people from great heights. According to the official rules, while rescuing people, the truck has to maintain a distance of at least 12 meters from the burning house.

Technical data of the engine:

Engine model:	Daimler Chrysler AG Econic 18/28 LL - Diesel
Construction year:	2004
Power:	205 kw (279 HP)
Cubic capacity:	6,374 cm³
Dimensions of engine:	Length 10 m, Width 2.5 m, Height 3.19 m
Dimensions of ladder:	Length 30 m
Weight of unloaded engine:	15,540 kg
Total weight:	18,000 kg

From what maximal height can the Munich fire brigade rescue people with this fire engine?

Figure 1. The modeling problem *fire brigade* (Blum & Borromeo Ferri, 2009).

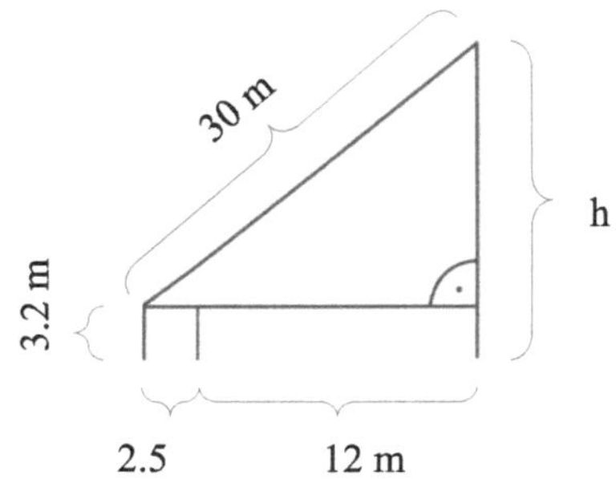

$$h \approx \sqrt{(30\,m)^2 - (12\,m + 2.5\,m)^2 + 3.2\,m}$$

$$\approx 29.5 \text{ m}$$

Answer: The fire brigade can rescue people who are at a height of about 30 m.

Figure 2. An exemplary solution to the *fire brigade* modeling problem.

the result by the scale. Either way, the student can obtain the mathematical result of 29.5 meters. Next, the mathematical result has to be *interpreted* with regard to reality. The student can interpret the result as the maximal rescue height when the engine is parallel-parked on the street. Finally, the solution and the chosen model need to be *validated*, that is, checked for plausibility and satisfaction. If the solution is considered unsatisfactory, the student might need to go through the modeling processing again by changing, for example, assumptions such as the parking position or the presence of a rescue cage. In addition, the student needs to plan, monitor, regulate, and evaluate their activities (Vorhölter, 2018) and to have beneficial attitudes and beliefs to work successfully through the modeling process (Maaß, 2006).

In recent decades, international research has documented deficits in students' modeling competencies. For example, on average in OECD countries, two thirds of fifteen-year-old students are not able to work on complex models of concrete situations that require students to, for example, make assumptions (OECD, 2016). Recent studies have investigated ways to help students develop their modeling competencies (e.g., by teacher interventions (Stender et al., 2017), or digital tools (Greefrath et al., 2018)). Another approach is to trigger the use of strategies that can offer powerful ways to apply the modeling process. A strategy that is considered particularly powerful in the domain of geometry is the strategy of self-generated drawing. Thus, triggering the construction of a drawing by explicitly asking students to make a drawing might be a way to promote students' task-specific modeling performance and their modeling competencies in the domain of geometry.

Self-generated drawing for mathematical modeling

The strategy of self-generated drawing in the context of solving a geometry modeling problem signifies the construction and use of a representative illustration to solve the modeling problem (Rellensmann et al., 2017). A drawing represents the geometry problem via visuo-spatial similarity. In a drawing, signs represent objects (e.g., a fire engine) that are arranged on the paper according to the relations given in the problem. To specify the relations between the objects, the student can integrate numbers (e.g., 3.19 to indicate a length), single variables (e.g., x to indicate unknown information), or words (e.g., *safety distance*) in the drawing.

A drawing that is made for a modeling problem can take two different forms (Presmeg, 2006; Rellensmann et al., 2017): A *situational drawing* refers to the model of the situation and shows the objects and relations pictorially, that is, according to their visual appearance (Figure 3). A *mathematical drawing* refers to the mathematical model and shows the problem structure mathematically, thereby focusing on the mathematical features of the problem (Figure 3).

Previous research from the domains of text comprehension and word problem solving has indicated that the construction and use of a situational or mathematical drawing might help students avoid or overcome barriers in the modeling process. As constructing a drawing can initiate cognitive and metacognitive processes by forcing the integration of multiple representations (Fiorella & Zhang, 2018), we suggest that it can also be beneficial for mathematical modeling.

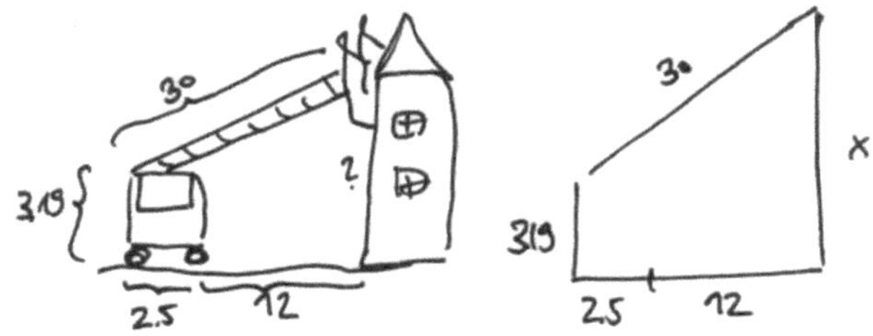

Figure 3. Exemplary situational (left) and mathematical (right) drawings.

Previous research has indicated that constructing a drawing can support text comprehension (Leopold & Leutner, 2012), help the problem solver see new information (e.g., mathematical objects and relations) (Cox, 1999), or initiate the metacognitive processes of self-monitoring and self-regulation (Stylianou, 2011). Theoretical analyses of students' solution processes and empirical findings from a case study have indicated that constructing a situational or mathematical drawing can initiate these processes and thereby support the solution of a modeling problem (Rellensmann, 2019).

Findings on the effectiveness of drawing instructions have been mixed and have indicated that whether or not students were previously trained to draw is important. For example, De Bock and colleagues (De Bock et al., 1998, 2003) and Bräuer et al. (2021) reported no or negative effects of drawing instructions on secondary school students' competencies in word problem solving and modeling in geometry without previous training, respectively. One explanation that previous studies have pointed to is that students made inaccurate drawings that were not helpful for solving the problems (Bräuer et al., 2021).

In previous studies that provided students with drawing training, students who were trained in drawing usually outperformed students without such training on a word problem solving test. For example, Csíkos et al. (2012) found that third-grade students who received 20 lessons on how to make drawings for word problems had higher achievement gains in arithmetic skills and word problem solving performance than their peers in the control condition. Van Essen and Hamaker (1990) reported higher achievement gains in word problem solving on the posttest for fifth-grade students after a 60-minute intervention. In both studies, drawing training was aimed at developing students' drawing construction skills and shaping their attitudes and beliefs with regard to drawing. Drawing training might have developed and extended students' strategic knowledge about the strategy of drawing (i.e., knowledge about different types of drawings and the characteristics of helpful drawings). Strategic knowledge about drawing is considered a precondition for the construction of accurate drawings, and accuracy is important for the effectiveness of drawing (Krawitz & Schukajlow, 2020; Rellensmann, 2019).

Drawing accuracy

Drawing accuracy describes the degree of correspondence between the problem structure described in the modeling problem and the problem structure that is represented in the drawing (Rellensmann, 2019; Van Essen & Hamaker, 1990). Examples of drawings with varying degrees of accuracy are presented in Figure 4.

A drawing with a high degree of accuracy (right drawing in Figure 4) adequately represents the modeling problem's relevant objects and relations, and the relevant numbers are correctly allocated in the drawing (Rellensmann et al., 2017). A drawing with an intermediate degree of accuracy (intermediate drawing in Figure 4) adequately represents the problems' objects and relations, but one or

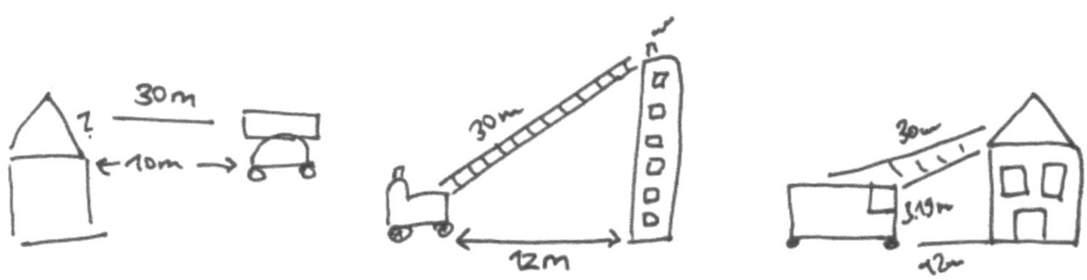

Figure 4. Student drawings incorrectly representing the relations (left), missing numerical information (middle), and accurately representing the problem structure (right).

more relevant numbers are missing. A drawing with a low degree of accuracy (left drawing in Figure 4) inadequately represents the problems' objects and relations, as relevant objects are missing or the objects that are drawn are incorrectly arranged on the paper.

The relation between drawing accuracy and modeling performance can be addressed on two levels: the task level and the student level. On the task level, non-aggregated task scores are used to investigate the relation between drawing accuracy and corresponding task performance. On the student level, aggregated task scores are used to investigate the relations between students' average drawing accuracy and average task performance across several tasks. Previous research has shown a positive relation between average drawing accuracy and modeling competencies in the domain of geometry on the student level (e.g., Rellensmann et al., 2017; Schukajlow et al., 2021). For example, Rellensmann et al. (2017) instructed students to make a situational drawing and a mathematical drawing for a geometry modeling problem and found relations of $r = .27$ and $r = .76$ between students' modeling competencies and their average situational and mathematical drawing accuracy, respectively. However, as the authors averaged drawing accuracy and modeling performance across several modeling problems, we do not know whether students who made a highly accurate drawing for a particular problem solved that particular modeling problem better. A case study of students' drawings and solutions indicated that accurate drawings were more likely to lead to correct solutions, whereas inaccurate drawings more often led to erroneous solutions under drawing instructions (Rellensmann, 2019). Also, findings from the case study suggested that the direction and strength of the task-level relation between drawing accuracy and modeling performance might vary between students (Rellensmann, 2019). One student might need an accurate drawing to solve a modeling problem, whereas another student can compensate for an inaccurate or incomplete drawing by keeping the specific details in his or her mind and can still solve the problem. Variation in the relation between drawing accuracy and modeling performance might be due to differences in mathematical abilities (Van Garderen et al., 2013) or working memory capacity (Taub et al., 2014).

As previously outlined, drawing accuracy can be considered crucial for the effectiveness of drawing. One approach that might help increase students' drawing accuracy involves promoting their strategic knowledge about drawing.

Strategic knowledge about drawing

According to the theoretical model by Borkowski et al. (1988), strategic knowledge about drawing is specific strategic knowledge. With regard to the strategy of self-generated drawing, it comprises knowledge about the characteristics of helpful situational and mathematical drawings that correspond to a given modeling problem and can help solve it (Rellensmann, 2019). For example, a student with good strategic knowledge about drawing knows that a helpful drawing accurately represents the problem structure and comprises all relevant numbers. Strategic knowledge makes up part of the knowledge component of metacognition (Brown, 1987; Flavell, 1979). It is distinct from the procedural component of metacognition, which comprises the actual use of a strategy (e.g., the construction and use of a drawing to solve a modeling problem) (Brown, 1987; Flavell, 1979). With regard to drawing instructions, Rellensmann (2019) found that strategic knowledge about situational and mathematical drawing predicted drawing accuracy and modeling competencies in the domain of geometry beyond the effects of cognitive abilities and interest in mathematics in grade 9. It can thus be concluded that strategic knowledge about situational and mathematical drawing is a precondition for fostering mathematical modeling, as both types of drawings can be helpful for solving a geometry modeling problem. However, the interplay between strategic knowledge, strategy use, and performance with respect to learner-generated drawing has rarely been addressed. The theoretical model by Borkowski et al. (2000) states that the effect of specific strategic knowledge on performance is transmitted via strategy use. With regard to self-generated drawing for modeling, the model would predict that strategic knowledge about drawing affects modeling competencies by means of drawing accuracy – an indicator of the quality of strategy use. Thus, we predict that drawing accuracy should mediate the effect of strategic knowledge about drawing on students' modeling competencies. Rellensmann et al. (2017) previously found a positive relation between strategic knowledge about drawing and modeling competencies with mathematical drawing accuracy as an intervening variable in the domain of geometry.

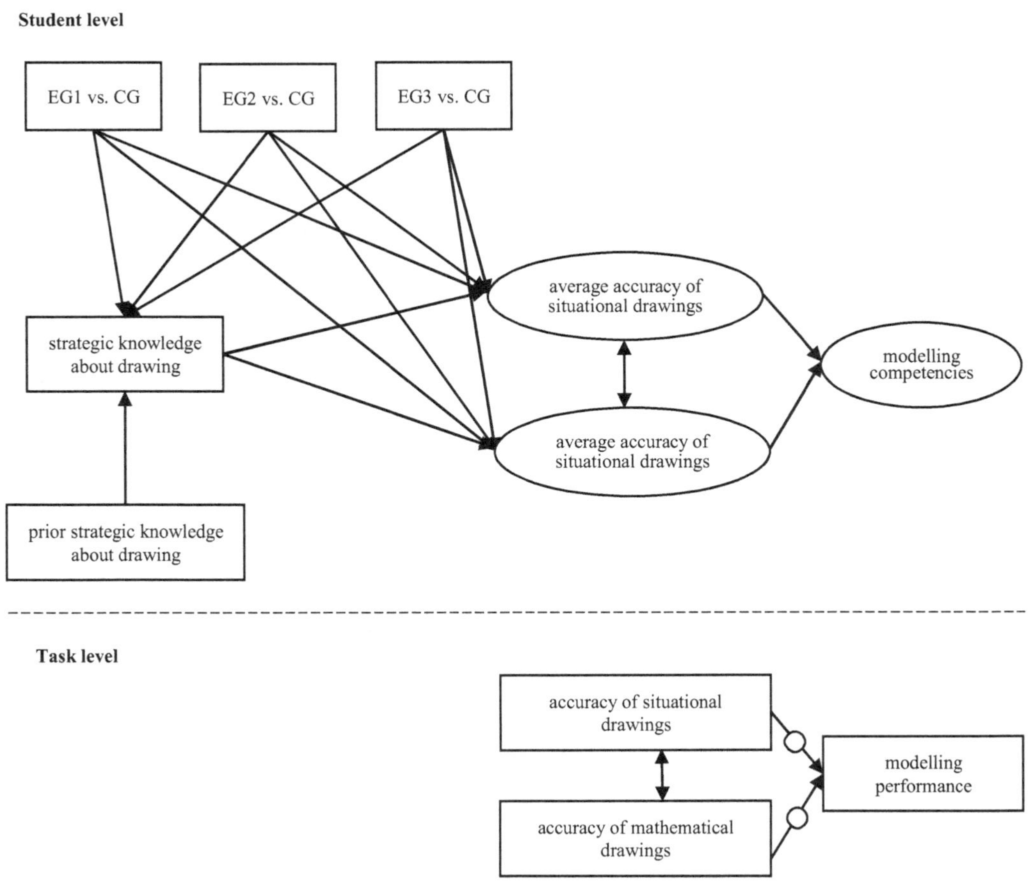

Figure 5. Illustration of the hypothesized two-level path model.

Research questions and hypotheses

In the present study, we hypothesized and tested a multilevel mediation model (Figure 5) to investigate whether and how strategic knowledge about drawing treatments affect students' modeling competencies. In addition, we analyzed how accuracy in situational and mathematical drawing is related to modeling competencies (i.e., average modeling performance across problems) and task-specific modeling performance.

By examining the hypothesized two-level path model (Figure 5), we investigated the following research questions and hypotheses, which we derived from the theory presented in the previous sections.

(1) How do treatments that were designed to improve students' strategic knowledge about situational drawing (EG1), mathematical drawing (EG2), or situational and mathematical drawing (EG3) affect students' strategic knowledge about drawing and their modeling competencies compared with a control group (CG)?

Strategic knowledge treatments will have a positive effect on students' strategic knowledge while controlling for students' prior strategic knowledge about drawing (before the treatments) (Hypothesis 1a). The treatment that was designed to improve students' strategic knowledge about situational and mathematical drawing will produce greater improvement in strategic knowledge about drawing than the treatments that were designed to improve strategic knowledge with regard to situational or mathematical drawing only (Hypothesis 1b). Strategic knowledge treatments will have a positive effect on students' modeling competencies (Hypothesis 1 c). The treatment that was designed to improve students' strategic knowledge about situational and mathematical drawing will produce greater improvements in modeling competencies than the treatments that were designed to improve strategic knowledge with regard to situational or mathematical drawing only (Hypothesis 1d). The effect of the treatments EG1, EG2, and EG3 on students' modeling competencies will be mediated by strategic knowledge about drawing and average accuracy in situational or mathematical drawing (Hypothesis 1e).

(1) How is accuracy in situational and mathematical drawings related to modeling competencies on the student level and to modeling performance on the task level?

On the student level, average accuracy in situational and mathematical drawings will be positively related to modeling competencies (Hypothesis 2a). On the task level, the accuracy of a situational or mathematical drawing will be positively related to modeling performance (Hypothesis 2b). The relation between situational or mathematical drawing accuracy and modeling performance will vary between students (Hypothesis 2 c). The assumed variation in the relation between situational or mathematical accuracy and modeling performance is modeled by random effects on the task level, which are illustrated by circles on the task-level paths in Figure 5.

Method

Participants

A total of 560 ninth-grade students from five middle track schools (German Realschulen) and one comprehensive school (German Gesamtschule) took part in the study. The 87 students from one middle track school were eliminated from the sample, as their mathematics teachers reported that the topic of the Pythagorean theorem had not been taught before the data were collected. This could readily be verified because the mean modeling performance of students attending this school was one standard deviation below the mean modeling performance in the other five schools. Of the remaining 473 students, 46.3% were female, and the mean age was 14.69 years.

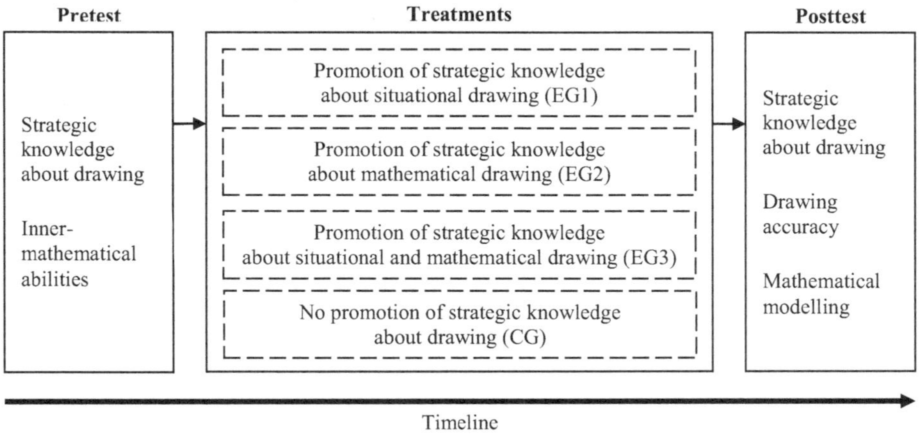

Figure 6. Overview of the study design.

Design

We employed a pre-posttest experimental between-subjects design (Figure 6). Students were randomly assigned to one of three treatment conditions, which were designed to promote strategic knowledge about situational drawing (EG1, n = 115), strategic knowledge about mathematical drawing (EG2, n = 124), strategic knowledge about situational and mathematical drawing (EG3, n = 125), or to the control condition in which we did not promote any knowledge about drawing (CG, n = 109). Each class was randomly divided into four groups with approximately the same number of students in each group. Each group was assigned to one of the three treatment conditions or the control condition. Students in the different conditions were provided with the respective training lesson in a separate room. The training lessons were led by master's students in accordance with the structure of a standardized manual. The master's students received an instruction, which covered 240 minutes in two days, before realizing the training lessons with students. To mitigate aggregate experimenter effects, all instructors performed the training lesson in each of the three treatment conditions and the control condition. As one school was eliminated from the sample in retrospect, the instructors performed at least three out of the four conditions in the final sample.

Procedure

With the pretest, we assessed students' strategic knowledge about drawing and their intra-mathematical abilities. The pretest was administered immediately before a two-week school holiday to minimize the effect of the strategic knowledge assessment (pretest) on the drawing-related measures (posttest). After the school holiday, students participated in one 90-minute standardized training lesson. Students' strategic knowledge about drawing was assessed with the posttest, which was administered three to four days after the training session. They were asked to solve modeling problems with instructions to make a drawing.

The training lessons

The treatments EG1, EG2, and EG3 covered one lesson of 90 minutes and were built on principles of effective strategy instruction from previous research (Dignath & Büttner, 2008). The main elements of the treatments were: instructor demonstration of strategy use, sensitization to the strategy's utility, explanation of strategy elements, demonstration of deficient strategy use, reflection of strategy use on different tasks, feedback on individual strategy use, and learning in varying social contexts.

The training lessons (EG1-3) began with the instructor's demonstration of how a situational or/and mathematical drawing could be constructed and used to solve an exemplary modeling problem. The instructor's demonstration served to define the respective type of drawing and to sensitize students to the utility of drawing strategies. After the demonstration, students were asked to individually reflect on the utility of the specific types of drawing in different phases of the modeling process and to share their thoughts with their partner. Second, students worked individually and afterward in pairs on characteristics of helpful drawings by comparing provided drawings of varying accuracy. The characteristics of drawing accuracy were then formalized and transferred to drawings for another modeling problem (Figure 7). In a classroom discussion, students discussed and corrected their responses. Third, students were asked to reflect on drawing accuracy by individually arranging given objects in an incomplete provided drawing to fit a given modeling problem and by comparing their drawing with their partner's drawing. Fourth, students were asked to compare and rank order provided drawings of varying accuracy. Students discussed and corrected their responses with their partner and in a classroom discussion.

In all three treatment conditions, the mathematical contents and tasks the students worked on were standardized. To ensure equal time on task, students in all conditions worked on the same number of drawings during the treatment, consisting of only situational drawings in EG1, only mathematical drawings in EG2, and half situational and half mathematical drawings in EG3. The mathematical contents of the treatments came from the domain of geometry (e.g., Intercept theorem), excluding the Pythagorean theorem and drawing to scale, which were mathematical topics that could be used to solve modeling problems on the posttest.

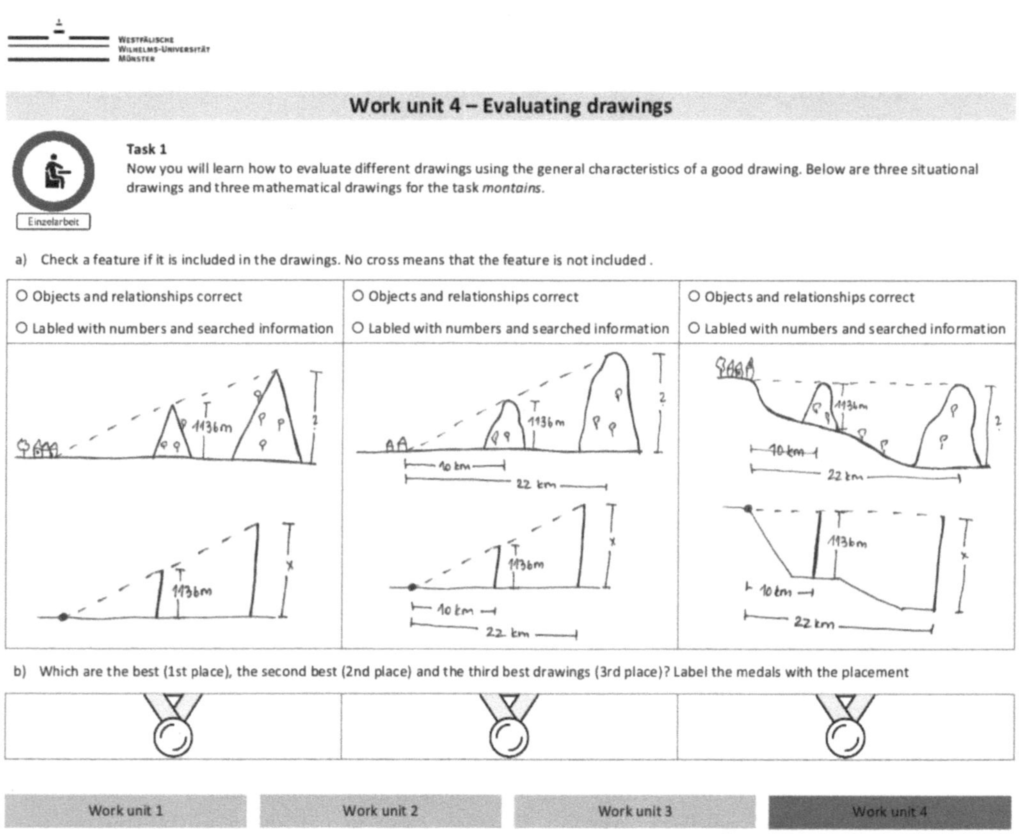

Figure 7. Instruction sheet for the treatment condition in which strategic knowledge about situational and mathematical drawing was promoted (EG3).

The control condition lesson followed the same chronology but was focused on the subject of symbolic term transformations rather than strategic knowledge. The lesson began with the instructor's demonstration of term transformations. Afterward, students worked individually and in pairs on four rules for term transformations (e.g., distributive law). In a classroom discussion, students discussed and corrected their responses. As students only worked on terms without exponents, the content of the control condition lesson was of little benefit for solving the modeling problems in the posttest.

Across the treatment conditions and the control condition, the social forms of learning were standardized, including individual work, pair work, and classroom discussions. For the pair work, the teacher grouped students from each part of a class into dyads (if it was not possible to group them into dyads only, then triads were also used) according to social compatibility and homogeneous achievement levels (in total, 224 dyads, 5 triads, and 10 students with missing data on partner assignment).

Measures

To assess strategic knowledge about drawing, intra-mathematical abilities, accuracy in situational and mathematical drawing, and modeling performance, we used well-evaluated instruments and scales from previous studies (Rellensmann et al., 2017, 2020).

Strategic knowledge about drawing

To assess students' strategic knowledge about drawing, we used the previously developed and validated strategic knowledge about drawing scale (Rellensmann et al., 2020). The strategic knowledge about drawing scale is constructed in such a manner that one real-world problem is followed by one item about situational drawings and one item about mathematical drawings (Figure 8). In the current study, we used 24 items from the scale. The items were split into a pretest and a posttest with eight identical items and eight items that differed between the pretest and posttest. Thus, on the pretest and posttest, students worked on 16 items, respectively. Half of the 16 items were about situational drawings, the other half of the items concerned mathematical drawings. To answer one item, students were asked to use a Likert scale to judge how helpful three provided drawings were for solving the real-world problem at hand. The provided drawings differed systematically with regard to their accuracy in representing the objects and relations and in the extent to which they completely represented the numbers. Students' answers on the Likert scale were transformed into a four-point scale representing the extent to which the rank order of drawings was correct. The maximum score (a code of 3) was given when the accurate and completely labeled drawing (first situational drawing in Figure 8) was judged to be more helpful than the correct but incompletely labeled drawing (third situational drawing in Figure 8) and the latter was judged to be more helpful than the inaccurate drawing (second situational drawing in Figure 8). The minimum score (a code of 0) was given for the inverse rank order, which meant that students judged the inaccurate drawing as the most helpful drawing.

We applied an item response theory approach to obtain student estimates of strategic knowledge about drawing. Data from the two measurement points were treated as two groups of independent "virtual" cases. In this way, estimates were scaled to the same scale, and differences could be interpreted in terms of individual change in repeated measures. All items met the mean-square criterion with $0.80 \leq wMNSQ \leq 1.13$ (Bond & Fox, 2015) and had a discrimination coefficient of $r \geq .36$. Further, the WLE reliability of .82 indicated that students' responses on the scale sufficiently corresponded with the expected values of the model. We exported weighted mean likelihood estimates (WLEs) of students' strategic knowledge about drawing to investigate the hypothesized multilevel mediation model.

Wind

The rotor blade of a 70 m high wind turbine has a length of 35 m and a so-called wind area of about 3,800 m^2. The wind area is the area that is swept by a rotor blade during a complete rotation. Can the information on the swept area be correct?

You do not have to solve the task!

First compare these three drawings with one another. Evaluate how useful each drawing is for solving the task.

Drawings	Not helpful at all	Not helpful	Rather not helpful	Rather helpful	Helpful	Very helpful
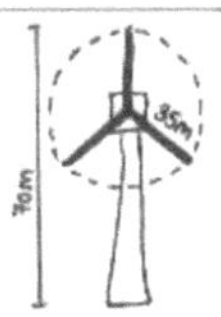	☐	☐	☐	☐	☐	☐
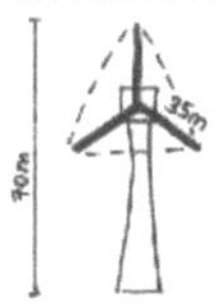	☐	☐	☐	☐	☐	☐
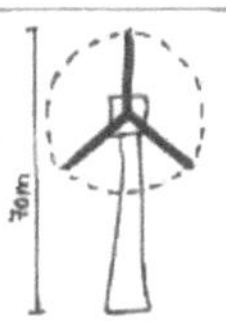	☐	☐	☐	☐	☐	☐

Next compare these three drawings with one another. Evaluate how useful each drawing is for solving the task.

Drawings	Not helpful at all	Not helpful	Rather not helpful	Rather helpful	Helpful	Very helpful
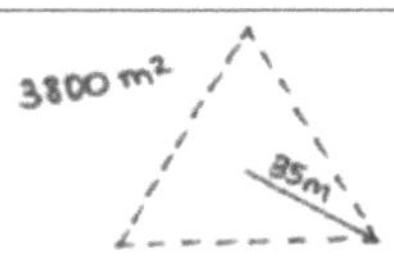	☐	☐	☐	☐	☐	☐
	☐	☐	☐	☐	☐	☐
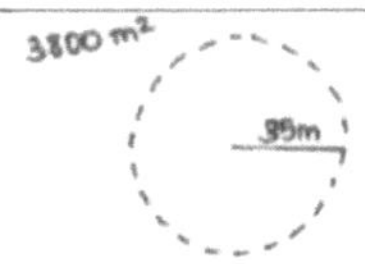	☐	☐	☐	☐	☐	☐

Figure 8. Exemplary wind item.

Intra-mathematical abilities

With the pretest, we also assessed students' intra-mathematical abilities with regard to the Pythagorean theorem (10 items). For example, students were asked to set up an equation that fit a provided right-angled triangle or to solve quadratic functions. Students' solutions were coded as correct (code 1) or incorrect (code 0). Cronbach's alpha was good ($\alpha = .73$).

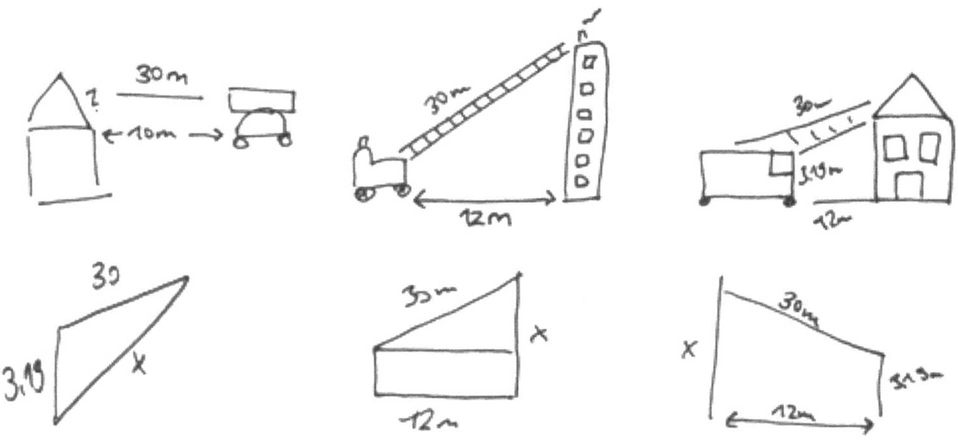

Figure 9. Exemplary situational and mathematical student drawings of varying accuracy (coded 0, 1, and 2 from left to right).

Accuracy of situational and mathematical drawings

Coding the accuracy of students' situational and mathematical drawings (6 items) was carried out in two steps. In the first step, a drawing was assigned to the situational or mathematical type of drawing (Figure 9). We computed the proportion of situationally drawn objects out of the total number of solution-relevant drawn objects. For a drawing in which the majority of the objects were drawn situationally, we assigned the drawing to the situational type of drawing. If the majority of the objects were drawn mathematically, we assigned the drawing to the mathematical type of drawing. Two experienced and well-trained students with bachelor's degrees in mathematics education coded the students' drawings and solutions in accordance with a detailed coding manual. Intercoder reliability (mean Cohen's κ = .83 across the six modeling problems and range from .74 to .92) indicated satisfactory agreement between coders with regard to a drawing's assignment to the situational or mathematical drawing type for a sample of 25% of the student drawings.

In the second step, the drawing's accuracy was rated (Figure 9). Full credit (a score of 2) was given to a drawing that correctly represented the objects and relations and in which all relevant numbers were inserted and correctly allocated. Partial credit (a score of 1) was given to a drawing in which the objects and relations were correctly represented but relevant numbers were missing or incorrectly allocated. A drawing that incorrectly represented the objects and relations was scored 0. A score of 0 was also given if no drawing had been made or if the drawing was of the other type of drawing. Intercoder reliability for drawing accuracy was satisfactory (mean Cohen's κ = .74 across the six modeling problems and range from .67 to .87). The scale's internal consistency was good with Cronbach's alpha values of .75 for situational drawing accuracy and .79 for mathematical drawing accuracy.

Modeling competencies

To assess students' modeling competencies, we used six geometry modeling problems, which required the students to determine a distance that could be modeled as one side of a right-angled triangle (Figure 1).

Students' task-specific modeling performance was scored on a three-point scale (6 items). We gave a score of 2 to a correct mathematical result based on a correct mathematical model (Figure 10). If the mathematical model was correct but computational errors resulted in an incorrect mathematical result, we gave a score of 1. A score of 0 was given to an incorrect or missing mathematical result based on an incorrect mathematical model. Intercoder reliability was satisfactory (mean Cohen's κ = .78 across the six modeling problems and range from .67 to .86). A Cronbach's alpha value of .74 indicated good internal consistency for the scale.

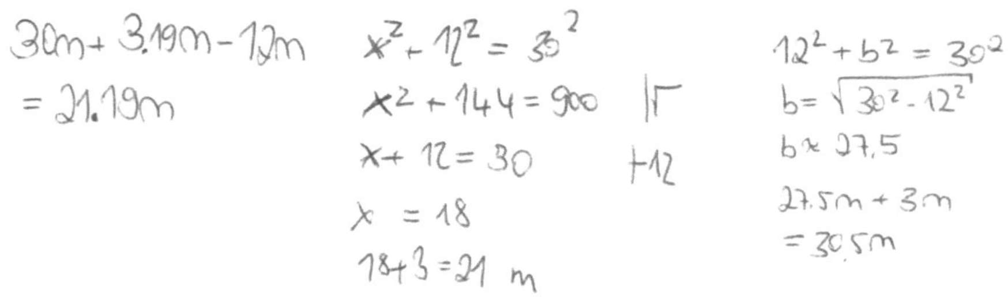

Figure 10. Students' mathematical models and calculations to the *fire brigade problem* (coded 0, 1, and 2 from left to right).

Data analysis

We used multilevel modeling to analyze our data. Multilevel modeling is an approach for analyzing data that have a nested structure (Raudenbush & Bryk, 2002). To investigate our research questions, we analyzed a total of 2,383 task scores (i.e., 2,383 scores for drawing accuracy and 2,382 scores for the corresponding task performance) that were nested in 473 students. Building a multilevel model allowed us to decompose the total variance of drawing accuracy and modeling performance into variance occurring within students (i.e., on the task level) and between students (i.e., on the student level). In this way, it was possible to simultaneously estimate effects on the task level (e.g., relation between drawing accuracy and modeling performance) and effects on the student level (e.g., effects of strategic knowledge about drawing on modeling competencies). Further, multilevel analysis gave us the ability to investigate how effects on the task level varied across students by specifying random effects.

For the analysis of the multilevel mediation model, we used the software MPlus (Version 7.31) and full likelihood estimation. Missing values were estimated by applying the Full Information Maximum Likelihood (FIML) algorithm. Our mediating variables on the task level were centered on their overall means (i.e., grand-mean-centered) to facilitate the interpretation of the estimates on level 2.

Results

Preliminary results

Students in the treatment conditions and the control condition did not differ in their intra-mathematical abilities ($F(3,435) < 1$), indicating that the randomized assignment of students to the treatment and control conditions was successful.

We further analyzed the number of situational and mathematical drawings in the different conditions to obtain evidence of the adequate implementation of the treatments. Across students and modeling problems on the posttest (on the task level), we found 25% situational drawings, 53% mathematical drawings, and 23% missing drawings. Students in EG1 who were provided with a lesson about situational drawings constructed more situational drawings and fewer mathematical drawings than students in the control condition ($\chi^2(1, N = 1344) = 9.48, p < .01$). Students in EG2 who were provided with a lesson about mathematical drawings constructed more mathematical drawings and fewer situational drawings than students in the control condition ($\chi^2(1, N = 1398) = 5.59, p < .05$). The use of situational and mathematical drawings by students in EG3 who were provided with a lesson about situational and mathematical drawings did not differ from those in the control condition ($\chi^2(1, N = 1404) = 1.07, p > .05$). These preliminary analyses provided us with evidence of the adequate implementation of the treatment, as students were more likely to construct the type of drawing that they had learned (EG1 and EG2) compared with the control condition, and students who learned about situational and mathematical drawings (EG3) were not biased toward either situational or mathematical types of drawings compared with the control condition. As students from all conditions were instructed to make a drawing, these findings indicate that students followed the drawing instructions on the posttest to a similar extent.

Descriptive statistics and correlations

Descriptive statistics and zero-order correlations are reported in Table 1. Except for the correlation between situational drawing accuracy and modeling competencies, all correlations were positive as expected. Negative correlations were found between situational and mathematical drawing accuracy on both levels, as students who made a situational drawing were given a score of 0 for mathematical drawing accuracy and vice versa.

Two-level path model

Analyses of variance showed that the majority of the variance (62%) in modeling performance was explained by task-level variation. The clustering within students (i.e., student-level variation) accounted for another 21% of the variance. These findings indicate the importance of considering variation at the task and student levels. By fitting the hypothesized model to the data, the amount of variance explained in modeling performance increased by 10% on the student level and by 30% on the task level compared with an unconditional model. Including random effects in the model explained 10% of the variance in task-level modeling performance compared with the same model with fixed effects.

Estimated regression coefficients for the direct effects on the student level are shown in Table 2. Indirect and total effects will be reported in the text. With regard to our hypotheses, we found a positive effect of the conditions EG1, EG2, and EG3 on students' strategic knowledge about drawing compared with the control condition (b_y = .44, .38, and .30, respectively, all *ps* < .01). That is, students in EG1, EG2, or EG3 had better strategic knowledge about drawing than students in the control condition while controlling for students' prior strategic knowledge. The value .44 for EG1 means that participating in EG1 compared with the control condition resulted in a mean increase in strategic knowledge about drawing of 0.44 of its standard deviation (i.e., 0.44 * 1.04 = 0.46). Thus, Hypothesis 1a was supported. However, participating in EG3 in which strategic knowledge about both situational and mathematical drawings was promoted did not result in higher strategic knowledge gains than the treatments that focused on only situational or mathematical drawings. Thus, Hypothesis 1b was not supported.

Table 1. Correlations, observed means, and standard deviations.

		Prior strategic knowledge (1)	Strategic knowledge (2)	Situational drawing accuracy (3)	Mathematical drawing accuracy (4)	Modeling competencies (5)
	1	1	.56**	.06	.28**	.20**
	2		1	.16**	.27**	.30**
	3			1	−.49**	−.04
	4			−.34**	1	.71**
	5			.11**	.46**	1
Total sample	*M* (*SD*)	0.81 (0.84)	0.97 (1.04)	0.50 (0.55)	0.89 (0.59)	0.63 (0.57)
EG1	*M* (*SD*)	0.73 (0.90)	0.91 (0.54)	0.48 (0.48)	0.58 (0.55)	0.91 (0.54)
EG2	*M* (*SD*)	0.83 (0.74)	1.06 (1.04)	0.24 (0.39)	1.18 (0.54)	0.61 (0.56)
EG3	*M* (*SD*)	0.76 (0.75)	0.96 (0.98)	0.51 (0.53)	0.93 (0.55)	0.70 (0.59)
CG	*M* (*SD*)	0.91 (0.97)	0.75 (1.10)	0.35 (0.47)	0.93 (0.56)	0.62 (0.57)

*p < .05, two-tailed, **p < .01, two-tailed. Student-level correlations are reported above the diagonal and task-level correlations are displayed below the diagonal.

Table 2. Estimated path coefficients for direct effects on the student level.

	Coefficients			*SE*
Paths	*b*	b_y	β	
From prior strategic knowledge to				
Strategic knowledge	0.71**		.57**	.08
From EG1 to				
Strategic knowledge	0.46**	.44**		.13
Accuracy in situational drawings	0.38**	.69**		.05
Accuracy in mathematical drawings	−0.38**	−.64**		.06
From EG2 to				
Strategic knowledge	0.39**	.38**		.11
Accuracy in situational drawings	−0.11**	−.20**		.04
Accuracy in mathematical drawings	0.09	.15		.07
From EG3 to				
Strategic knowledge	0.31**	.30**		.11
Accuracy in situational drawings	0.12**	.22**		.05
Accuracy in mathematical drawings	−0.04	.07		.07
From strategic knowledge to				
Accuracy in situational drawings	0.04**		.08**	.02
Accuracy in mathematical drawings	0.11**		.19**	.02
From accuracy in situational drawings to				
Modeling competencies	0.07		.07	.05
From accuracy in mathematical drawings to				
Modeling competencies	0.20**		.21**	.05

*$p \leq .05$, **$p < .01$, one-tailed. *b* = unstandardized regression coefficient, b_y = partially standardized regression coefficient with respect to the standard deviation of the dependent variable, β = fully standardized regression coefficient, *SE* = standard error

Contrary to our expectations, we did not find a total effect of our treatment conditions on students' modeling competencies (student level) compared with the control condition. This means that, in EG1, EG2, and EG3, students' strategic knowledge about drawing improved, but these students did not demonstrate higher modeling competencies than students in the control condition. Thus, Hypothesis 1 c was not supported. Consequently, the treatment in which strategic knowledge about situational and mathematical drawings was promoted did not result in higher modeling competencies than the treatments that focused on only situational or mathematical drawings. Hypothesis 1d was also not supported.

As expected, we found positive indirect effects of EG1, EG2, and EG3 on modeling competencies via gains in strategic knowledge about drawing and average situational drawing accuracy (b_y = .01, .01, and .01, all *ps* < .05) and average mathematical drawing accuracy (b_y = .04, .03, and .02, all *ps* < .01). That is, the strategic knowledge about drawing of students who participated in EG1, EG2, or EG3 improved, and, as a result, these students constructed on average more accurate situational and mathematical drawings and demonstrated higher modeling competencies than their peers in the control condition. Thus, Hypothesis 1e, which proposed a mediated effect, was confirmed.

On the student level, the expected positive effect of accuracy in situational drawings on modeling competencies was at the edge of significance (β = .07, $p = .08$). Consistent with our expectations, accuracy in mathematical drawings had a positive effect on modeling competencies on the student level (β = .21, $p < .01$). That is, students who constructed more accurate mathematical drawings on average demonstrated higher modeling competencies than their peers who constructed less accurate drawings on average. Thus, Hypothesis 2a was supported.

On the task level, accuracy in situational drawings and accuracy in mathematical drawings had an average positive effect on modeling performance (*b* = .38 and .53, respectively, both *ps* < .01). That is, a student who made a situational or mathematical drawing of high quality solved the respective modeling problem better. However, as expected, the effects varied between students (.11, and .11, *ps* < .01) in terms of the strengths of the effects (the 95% confidence intervals were [.31;.45] and [.47;.58], respectively). This means that for some students, the positive relation between situational and mathematical drawing accuracy and task-specific modeling performance was stronger than for other students. Thus, Hypotheses 2b and 2c were supported.

Discussion

We conducted an experimental study to investigate the effects of students' strategic knowledge about drawing on their modeling competencies. To specify the mechanisms underlying the investigated effects, we conducted a multilevel mediation analysis with accuracy in situational and mathematical drawings as intervening variables.

Effects of strategic knowledge about drawing on modeling competencies

With regard to students' strategic knowledge about drawing, we found that the strategic knowledge treatments improved students' strategic knowledge about drawing. However, in the current study, the promotion of strategic knowledge about drawing did not result in an increase in students' modeling competencies. One explanation is that the treatment duration of 90 minutes was sufficient to develop and expand strategic knowledge about drawing, but students needed further training to transfer their strategic knowledge about drawing into action when solving unknown geometry modeling problems with drawing instructions. This explanation was supported by Van Essen and Hamaker (1990), who reported that students did not benefit from drawing for problems whose problem structure was different from the ones used in the 60-minute treatment. In turn, further practice in transferring strategic knowledge about drawing into accurate drawings might help students consolidate and even expand their strategic knowledge about drawing.

As expected, the results indicated an indirect effect: Students who improved their strategic knowledge about drawing constructed on average more accurate situational and mathematical drawings and demonstrated higher modeling competencies than their peers in the control condition. Thus, in line with previous research on strategic knowledge (Schneider & Artelt, 2010), we found that strategic knowledge about drawing was a necessary but not sufficient pre-condition for the effective use of drawings. As we found an indirect effect of strategic knowledge about drawing on modeling competencies with average drawing accuracy as an intervening variable, the chain of effects hypothesized in the theoretical model by Borkowski et al. (2000) was confirmed for the drawing strategy and modeling in the domain of geometry in the present study. This new finding has theoretical implications for research on strategies, as it confirms the importance of strategic knowledge and high-quality strategy use for students' achievement-related outcomes such as modeling competencies. Future research should investigate whether this chain of effects holds for other indicators of high-quality drawing use (e.g., drawing coherence, which describes the extent to which a student's visual and symbolic representations match, or a drawing's focus on relevant information), other strategies (e.g., underlining), other mathematical topics (e.g., functions), and other domains (e.g., science).

In the current study, improving students' strategic knowledge about drawing in isolation was not an immediate way to improve students' modeling competencies. Considering the indirect effect found in the current study and the theoretical model by Borkowski et al. (2000), training students to construct and use drawings (i.e., strategy use) might be a second crucial component of drawing training to improve modeling competencies (Csíkos et al., 2012; Uesaka et al., 2010; Van Essen & Hamaker, 1990). Previous research generated overviews of how students can be trained to construct drawings (Fiorella & Zhang, 2018). For example, students can be asked to translate short passages from modeling problem texts into drawings, to complete partially incomplete drawings that are given, or to compare a self-constructed drawing with a highly accurate drawing and to revise one's own drawing before solving the modeling problem. Future research should investigate how isolated and combined training in strategic knowledge and strategy use (i.e., the construction and use of a drawing) contribute to students' abilities to construct accurate drawings and to their modeling competencies when they are instructed to make a drawing. In addition, shaping students' motivation, positive emotions, and beliefs with regard to drawing strategies might be central for their use of drawing strategies to solve modeling problems (Schukajlow et al., 2021) and for the effectiveness of drawing training (Uesaka et al., 2010). This may be the case, for example, because beliefs about the usefulness of drawing might affect students' acquisition of strategic knowledge about drawing and their engagement in drawing. Teachers should encourage students to make drawings and highlight how drawings can help students successfully solve a modeling problem, thereby shaping, for example, students' perceptions of utility and self-efficacy in drawing.

In the present study, students who learned about situational and mathematical drawing did not show a greater gain in strategic knowledge about drawing nor did they have higher modeling competencies than students who learned about only one type of drawing. Understanding situational and mathematical drawing as two drawing strategies for mathematical modeling, this finding is in line with a meta-analysis by Hembree (1992), which showed that teaching students one strategy yielded a larger effect than teaching them several strategies. One explanation for the current study's finding of no difference in strategic knowledge between EG3 and either EG1 or EG2 is that students who learned about situational and mathematical drawing spent less time with one type of drawing and developed less profound strategic knowledge than students who learned about situational or mathematical drawing in depth. Another explanation is that learning about situational and mathematical drawing at the same time results in cognitive overload, which prevents students from benefitting from the combined strategic knowledge treatment. Further research is needed to figure out why learning about situational and mathematical drawings does not yield greater strategic knowledge gains than learning about one drawing type only. Mathematics instruction that is aimed at developing students' modeling competencies should introduce students to both types of drawing, as the current study's findings showed that situational and mathematical drawing are powerful strategies that can support the modeling process. The current study's findings suggest that students' strategic knowledge about situational and mathematical drawing should be promoted one after the other to avoid cognitive overload.

Relation between drawing accuracy and modeling competencies

In line with previous research (e.g., Rellensmann et al., 2017), we found a positive correlation between average drawing accuracy and students' modeling competencies for mathematical drawings and–at the edge of significance–for situational drawings on the student level. This finding indicates that students who are better at constructing accurate situational or mathematical drawings have better modeling competencies compared with students who are less skilled in constructing drawings. Thus, the ability to construct accurate situational and mathematical drawings is important for students' modeling competencies in the domain of geometry. The weak relation between average situational drawing accuracy and modeling competencies can be explained by the interference of students' mathematics achievement: Students with lower achievement levels make more frequent use of situational drawings (Van Garderen et al., 2013) but lack modeling competencies, for example, with regard to mathematizing, compared with students with higher achievement levels. Future research should address the reasons for students' decision to make a situational or mathematic drawing while solving a modeling problem.

On the task level, both situational and mathematical drawing accuracy were positively related to task-specific modeling performance. If a more accurate situational or mathematical drawing was made, the particular modeling problem was solved better. This finding is consistent with the theory of modeling, as it confirms that situational drawings and mathematical drawings are potentially powerful types of drawings that can help solve a geometry modeling problem (Rellensmann, 2019). This result is important because it highlights and confirms the direct relation between a student's representation of a mathematical task and the student's modeling performance in this specific task as described in theoretical descriptions of modeling processes (e.g., Blum & Leiss, 2007; Galbraith &

Stillman, 2006) and as highlighted by qualitative results (Rellensmann, 2019). Future research should investigate differential effects of situational and mathematical drawings on modeling performance and possible interactions with student and task characteristics, as a case study provided us with the hypothesis that situational and mathematical drawings can support different phases in a modeling process and might depend on, for example, students' intra-mathematical abilities (Rellensmann, 2019). On the basis of current knowledge, mathematics teachers should introduce students to both types of drawing (e.g., Presmeg, 2006). For example, the teacher can demonstrate the construction of a situational and a mathematical drawing to solve a modeling problem and show how the drawings can help solve a modeling problem. The decision to make a situational or mathematical drawing for a specific modeling problem can be left up to the individual student.

Further, in the present study, the relation between situational and mathematical drawing accuracy and modeling performance varied between students. More precisely, the positive relation was stronger for some students than for others. One explanation refers to students' intra-mathematical abilities (Rellensmann, 2019). Students who lack intra-mathematical abilities can be expected to benefit less from constructing a drawing, as they might construct an accurate drawing but fail to recognize that they can apply the Pythagorean theorem. Thus, one practical implication might be that teachers should ensure that students are familiar with the underlying mathematics to make students experience drawing as a helpful strategy for solving a modeling problem.

Strengths and limitations

In the current research, we investigated the effects of training strategic knowledge about drawing on the accuracy of different types of drawings and on modeling competencies. We assessed students' modeling competencies in the domain of geometry, as drawing for modeling problems is expected to be particularly effective in this domain (Hershkowitz et al., 1989). Previous research has indicated that effects of strategic knowledge might depend on the specific mathematical domain. For example, Lucangeli and Cornoldi (1997) found a stronger relation between metacognition and performance in the domain of geometry than in arithmetic. In future research, the role of strategic knowledge should be addressed in other mathematical domains (e.g., linear functions) to help generalize findings on strategic knowledge about drawing. Also, the results of the current research are restricted to modeling problems like the ones used in this study. In the future, the effects of strategic knowledge and the role of drawing for solving more complex modeling problems should be investigated to generalize results across different types of modeling problems. Also, the measure of modeling competencies in the current study was limited to an analysis of students' written answers, which cannot capture the complexity of the modeling process. Further studies with fine-grained measures of modeling competencies are needed to deepen our understanding of the relations between strategic knowledge, drawing accuracy, and modeling activities. With regard to the measure of strategic knowledge about drawing, we focused on two characteristics that are central to helpful drawings, i.e., the correct representation of relevant objects and relations and the correct and complete labeling with numbers. Other facets of strategic knowledge (e.g., focusing on solution-relevant information) should be added to the measure in future research to develop a more comprehensive measure of strategic knowledge. Also, the relations between strategic knowledge, strategy use, and modeling competencies should be confirmed for indicators of high-quality drawing use other than drawing accuracy (e.g., a focus on solution-relevant information). Finally, the hypothesized and tested multilevel mediation model might be incomplete and could be improved by adding additional predictors. In future studies, the model could be expanded by adding, for example, students' intra-mathematical abilities or spatial abilities, as previous research has indicated that these affect the effects of drawing on performance (Fiorella & Zhang, 2018; Kozhevnikov et al., 2002). Another limitation concerns the coding of drawing accuracy and the interpretation of findings: The observed negative correlations between situational and mathematical drawing accuracy result from the applied coding scheme, which allowed us to investigate indirect effects of our interventions on modeling competencies via drawing accuracy. To specifically investigate the relation between situational and mathematical drawing accuracy a different research design and another coding approach is needed. Another limitation concerns the duration of the treatments in the current study. Although the short-term treatments were sufficient for detecting effects of strategic knowledge about drawing on modeling competencies, larger and more sustainable effects can be expected for longer and more comprehensive treatments that combine strategic knowledge training with strategy use. Whether the effects of strategic knowledge on modeling competencies are sustainable should be investigated in future research.

Disclosure statement

No potential conflict of interest was reported by the author(s).

Funding

The present study was conducted as part of the project *Visualization while solving modelling problems*, which has been funded by the German Research Foundation (Deutsche Forschungsgemeinschaft, grant number SCHU 2629/3-1) and by the Swiss National Science Foundation (Schweizer Nationalfond, grant number 100019E-164816/1).

ORCID

Johanna Schoenherr (née Rellensmann) http://orcid.org/0000-0002-6445-2917

References

Blomhøj, M., & Jensen, T. H. (2003). Developing mathematical modelling competence: Conceptual clarification and educational planning. *Teaching Mathematics and Its Applications*, *22*(3), 123–139. https://doi.org/10.1093/teamat/22.3.123

Blum, W., & Borromeo Ferri, R. (2009). Mathematical modelling: Can it be taught and learnt? *Journal of Mathematical Modelling and Application*, *1*(1), 45–58.

Blum, W., & Leiss, D. (2007). How do students and teachers deal with mathematical modelling problems? The example sugarloaf and the DISUM project. In C. Haines, P. L. Galbraith, W. Blum, & S. Khan (Eds.), *Mathematical modelling (ICTMA 12): Education, engineering and economics* (pp. 222–231). Horwood.

Bond, T., & Fox, C. M. (2015). *Applying the Rasch model. Fundamental measurement in the human sciences*. Routledge.

Borkowski, J. G., Chan, L. K., & Muthukrishna, N. (2000). A process-oriented model of metacognition: Links between motivation and executive functioning. In G. Schraw & J. C. Impara (Eds.), *Issues in the measurement of metacognition* (pp. 1–41). University of Nebraska Press.

Borkowski, J. G., Milstead, M., & Hale, C. (1988). Components of children's metamemory: Implications for strategy generalization. In F. E. Weinert & M. Perlmutter (Eds.), *Memory development: Universal changes and individual differences* (pp. 73–100). Lawrence Erlbaum.

Bräuer, V., Leiss, D., & Schukajlow, S. (2021). Skizzen zeichnen zu Modellierungsaufgaben - Eine Analyse themenspezifischer Differenzen einer Visualisierungsstrategie beim mathematischen Modellieren [Drawing for modelling problems. An analysis of content-specific differences of a visualazation strategy to solve modelling problems]. *Journal für Mathematik-Didaktik* 42 , 1–33 doi:10.1007/s13138-021-00182-7.

Brown, A. L. (1987). Metacognition, executive control, self-regulation, and other mysterious mechanisms. In F. E. Weinert & R. H. Kluwe (Eds.), *Metacognition, motivation, and understanding* (pp. 65–116). Erlbaum.

Cox, R. (1999). Representation construction, externalised cognition and individual differences. *Learning and Instruction*, *9*(4), 343–363. https://doi.org/10.1016/S0959-4752(98)00051-6

Csíkos, C., Szitányi, J., & Kelemen, R. (2012). The effects of using drawings in developing young children's mathematical word problem solving: A design experiment with third-grade Hungarian students. *Educational Studies in Mathematics*, *81*(1), 47–65. https://doi.org/10.1007/s10649-011-9360-z

De Bock, D., Verschaffel, L., Janssens, D., Van Dooren, W., & Claes, K. (2003). Do realistic contexts and graphical representations always have a beneficial impact on students' performance? Negative evidence from a study on modeling non-linear geometry problems. *Learning and Instruction*, *13*(4), 441–463. https://doi.org/10.1016/S0959-4752(02)00040-3

De Bock, D., Verschaffel, L., & Janssens, D. (1998). The predominance of the linear model in secondary school students' solutions of word problems involving length and area of similar plane figures. *Educational Studies in Mathematics*, *35* (1), 65–83. https://doi.org/10.1023/A:1003151011999

Dignath, C., & Büttner, G. (2008). Components of fostering self-regulated learning among students. A meta-analysis on intervention studies at primary and secondary school level. *Metacognition and Learning*, *3*(3), 231–264. https://doi.org/10.1007/s11409-008-9029-x

English, L., & Sriraman, B. (2010). Problem solving for the 21st century. In B. Sriraman & L. English (Eds.), *Theories of mathematics education: Seeking new frontiers* (pp. 263–290). Springer.

Fiorella, L., & Zhang, Q. (2018). Drawing boundary conditions for learning by drawing. *Educational Psychology Review*, *30*(3), 1115–1137. https://doi.org/10.1007/s10648-018-9444-8

Flavell, J. H. (1979). Metacognition and cognitive monitoring: A new area of cognitive developmental inquiry. *American Psychologist*, *34*(10), 906–911. https://doi.org/10.1037/0003-066X.34.10.906

Galbraith, P. L., & Stillman, G. (2006). A framework for identifying student blockages during transitions in the modelling process. *ZDM–The International Journal on Mathematics Education*, *38*(2), 143–162. https://doi.org/10.1007/BF02655886

Geiger, V., & Frejd, P. (2015). A reflection on mathematical modelling and applications as a field of research: Theoretical orientation and diversity. In G. A. Stillman, W. Blum, & M. S. Biembengut (Eds.), *Mathematical modelling in education research and practice* (pp. 161–171). Springer.

Greefrath, G., Hertleif, C., & Siller, H.-S. (2018). Mathematical modelling with digital tools—a quantitative study on mathematising with dynamic geometry software. *ZDM*, *50*(1), 233–244. https://doi.org/10.1007/s11858-018-0924-6

Hembree, R. (1992). Experiments and relational studies in problem solving: A meta-analysis. *Journal for Research in Mathematics Education*, *23*(3), 242–273. https://doi.org/10.2307/749120

Hershkowitz, R., Ben-Chaim, D., Hoyles, C., Lappan, G., Mitchelmore, M., & Vinner, S. (1989). Psychological aspects of learning geometry. In P. Nesher & J. Kilpatrick (Eds.), *Mathematics and cognition (ICMI study series)* (pp. 70–95). University Press.

Kaiser, G., & Sriraman, B. (2006). A global survey of international perspectives on modelling in mathematics education. *Zentralblatt Für Didaktik der Mathematik*, *38*(3), 302–310. https://doi.org/10.1007/BF02652813

Klieme, E., Hartig, J., & Rauch, D. (2008). The concept of competence in educational contexts. In J. Hartig, E. Klieme, & D. Leutner (Eds.), *Assessment of competencies in educational contexts* (pp. 3–22). Hogrefe & Huber.

Kozhevnikov, M., Hegarty, M., & Mayer, R. (2002). Revising the visualizer-verbalizer dimension: Evidence for two types of visualizers. *Cognition and Instruction*, *20*(1), 47–77. https://doi.org/10.1207/S1532690XCI2001_3

Krawitz, J., & Schukajlow, S. (2020). When can making a drawing hinder problem solving? Effect of the drawing strategy on linear overgeneralizations and problem solving. *Frontiers in Psychology*, *11*, 506. https://doi.org/10.3389/fpsyg.2020.00506

Leopold, C., & Leutner, D. (2012). Science text comprehension: Drawing, main idea selection, and summarizing as learning strategies. *Learning and Instruction*, *22*(1), 16–26. https://doi.org/10.1016/j.learninstruc.2011.05.005

Lucangeli, D., & Cornoldi, C. (1997). Mathematics and metacognition: What is the nature of the relationship? *Mathematical Cognition*, *3*(2), 121–139. https://doi.org/10.1080/135467997387443

Maaß, K. (2006). What are modelling competencies? *ZDM–The International Journal on Mathematics Education*, *38*(2), 113–142. https://doi.org/10.1007/BF02655885

Niss, M., Blum, W., & Galbraith, P. L. (2007). Introduction. In W. Blum, P. L. Galbraith, H.-W. Henn, & M. Niss (Eds.), *Modelling and applications in mathematics education: The 14th ICMI study* (pp. 1–32). Springer.

OECD. (2016). *Mathematics performance among 15-year-olds PISA 2015 results (Volume I): Excellence and equity in education.*

Presmeg, N. C. (2006). Research on visualization in learning and teaching mathematics. In A. Gutiérrez & P. Boero (Eds.), *Handbook of research on the psychology of mathematics education* (pp. 205–235). Sense Publishers.

Raudenbush, S., & Bryk, A. (2002). *Hierarchical linear models: Applications and data analysis methods.* Sage Publications.

Rellensmann, J. (2019). Selbst erstellte Skizzen beim mathematischen Modellieren: Ergebnisse einer empirischen Untersuchung [Using self-generated drawings to solve modelling problems: Results of an empirical study]. Wiesbaden: Springer Spektrum.

Rellensmann, J., Schukajlow, S., & Leopold, C. (2017). Make a drawing. Efects of strategic knowledge, drawing accuracy, and type of drawing on students' mathematical modelling performance. *Educational Studies in Mathematics, 95*(1), 53–78. https://doi.org/10.1007/s10649-016-9736-1

Rellensmann, J., Schukajlow, S., & Leopold, C. (2020). Measuring and investigating strategic knowledge about drawing to solve geometry modelling problems. *ZDM, 52*(1), 97–110. https://doi.org/10.1007/s11858-019-01085-1

Schneider, W., & Artelt, C. (2010). Metacognition and mathematics education. *ZDM–The International Journal on Mathematics Education, 42*(2), 149–161. https://doi.org/10.1007/s11858-010-0240-2

Schukajlow, S., Blomberg, J., Rellensmann, J., & Leopold, C. (2021). The role of strategy-based motivation in mathematical problem solving: The case of learner-generated drawings. *Learning and Instruction, Article* 101561. https://doi.org/10.1016/j.learninstruc.2021.101561

Schukajlow, S., Blomberg, J., Rellensmann, J., & Leopold, C. (2021). Do emotions and prior performance facilitate the use of the learner-generated drawing strategy? Effects of enjoyment, anxiety, and intramathematical performance on the use of the drawing strategy and modelling performance. *Contemporary Educational Psychology, 65,* 101967. https://doi.org/10.1016/j.cedpsych.2021.101967

Stender, P., Krosanke, N., & Kaiser, G. (2017). Scaffolding complex modelling processes: An in-depth study. In: Stillman, G., Blum, W., and Kaiser, G. *Mathematical modelling and applications: Crossing and researching boundaries in mathematics education* (pp. 467–477). Cham: Springer.

Stillman, G., Blum, W., & Kaiser, G. (2017). *Mathematical modelling and applications: Crossing and researching boundaries in mathematics education.* Springer.

Stylianou, D. A. (2011). An examination of middle school students' representation practices in mathematical problem solving through the lens of expert work: Towards an organizing scheme. *Educational Studies in Mathematics, 76*(3), 265–280. https://doi.org/10.1007/s10649-010-9273-2

Taub, M., Azevedo, R., Bouchet, F., & Khosravifar, B. (2014). Can the use of cognitive and metacognitive self-regulated learning strategies be predicted by learners' levels of prior knowledge in hypermedia-learning environments?. *Computers in Human Behavior, 39,* 356–367 doi:https://doi.org/10.1016/j.chb.2014.07.018.

Uesaka, Y., Manalo, E., & Ichikawa, S. (2010). The effects of perception of efficacy and diagram construction skills on students' spontaneous use of diagrams when solving math word problems. In A. K. Goel, M. Jamnik, & N. H. Narayanan (Eds.), *Diagrammatic representation and inference* (Vol. 6170, pp. 197–211). Springer.

Van Essen, G., & Hamaker, C. (1990). Using self-generated drawings to solve arithmetic word problems. *The Journal of Educational Research, 83*(6), 301–312. https://doi.org/10.1080/00220671.1990.10885976

Van Garderen, D., Scheuermann, A., & Jackson, C. (2013). Examining how students with diverse abilities use diagrams to solve mathematics word problems. *Learning Disability Quarterly, 36*(3), 145–160. https://doi.org/10.1177/0731948712438558

Vorhölter, K. (2018). Conceptualization and measuring of metacognitive modelling competencies: Empirical verification of theoretical assumptions. *ZDM Mathematics Education, 50*(1), 343–354. https://doi.org/10.1007/s11858-017-0909-x

Metacognition in Mathematical Modelling: the Connection between Metacognitive Individual Strategies, Metacognitive Group Strategies and Modelling Competencies

Katrin Vorhölter

ABSTRACT

There are several conceptualizations of modeling competencies, including among others on the one hand so-called sub-competencies, which are required to progress from one step of a modeling process to the next, and metacognitive individual and group strategies. However, the relationship between metacognitive strategies and modeling sub-competencies remains unclear, as does the influence of metacognitive strategies on the development of modeling competencies. The current paper presents the results of a study conducted with 170 students in grades nine and ten. This intervention study concerns the relationships between metacognitive strategies and the development of these strategies and modeling sub-competencies over the course of the study. The results illustrate that metacognitive individual strategies are highly correlated with each other, and metacognitive group strategies are highly correlated with each other, but metacognitive individual strategies are not correlated with metacognitive group strategies. Analysis of the development of metacognitive strategies and modeling sub-competencies within the intervention study reveals that not all of these developed as expected. Additionally, the factors of metacognitive strategies that were measured within the study allow for limited conclusions about the development of students' modeling sub-competencies. These results have implications for further research.

Introduction

Different kinds of competencies are required to develop solutions to modeling problems. For one, there are so-called sub-competencies of modeling, which relate to the sub-steps of a modeling process. Furthermore, global competencies, such as social and communication competencies, are needed to go through the individual steps of a modeling process in a goal-oriented way, to work effectively (in small groups), and to present and communicate the obtained results. Metacognition is considered to be one of these global competencies that is useful not only for working on modeling problems but also for developing modeling competencies (e.g., Kaiser & Brand, 2015).

Given the assumed importance of different competencies for successfully working on a modeling problem, there remains "a particularly challenging open research question to establish a theoretically and empirically based competence model for mathematical modelling" (Blum, 2011, 21). Extensive research is required to develop such a competence model. While several studies have investigated the development of sub-competencies for modeling, relatively few studies have focused on metacognition in modeling so far.

One reason for this may be that metacognition is a relatively vague concept that was developed in different domains with different conceptualizations. A distinction is typically drawn between metacognitive knowledge and metacognitive strategies. In addition, the use of metacognitive strategies at the individual level is distinguished from their use at the group level. When small groups work on modeling problems, it is important not only for metacognitive strategies to be used by individuals but also for metacognitive strategies to be used to regulate the group's working process.

In this paper, the theoretically assumed relationship between metacognitive individual and group strategies, the development of metacognitive strategies as well as the development of modeling sub-competencies are empirically analyzed.

Theoretical framework

Scholars have repeatedly asserted the importance of metacognition for successful modeling processes and for the development of modeling competencies (Blum, 2015). To properly analyze the relationship between metacognitive individual and group strategies and their impact on the development of modeling competencies, researchers must theoretically and empirically consider essential characteristics. However, there are only a few detailed theoretical conceptualizations of this relationship, and as a consequence, few studies have empirically examined the relationship. The following section presents existing theoretical and empirical considerations.

Mathematical modeling competencies

The process of working on a modeling problem is usually represented in the form of a modeling cycle. Depending on the perspective and purpose of modeling (Kaiser, 2017), various representations have been developed with different degrees of differentiation (e.g., Kaiser, 2017). What they all have in common is that a real situation (also called a real problem) represents the starting point of a modeling process. This situation must be understood, simplified, and structured to arrive at a representation of reality in the form of a real model. This real model becomes a mathematical model via translation into mathematical language, and by working on it, one arrives at a mathematical solution. This solution is then interpreted and transformed into a real solution, which must be validated with respect to both the real problem and the real model. Validation of the real solution may require repeating the cycle. In practice, the steps are usually not processed chronologically. Rather, there are so-called "mini-cycles," in which one moves between steps (Borromeo Ferri, 2018). *Understanding and simplifying* the real problem, *mathematizing* the real model into a mathematical model, *working mathematically* in this model,[1] *interpreting* the results for the real situation, *validating* the results, and *exposing* them are, in the current discourse, referred to as sub-competencies of modeling (e.g., Greefrath & Vorhölter, 2016) as one facet of modeling competencies.

In addition, in order to work successfully on modeling problems, there must be global modeling competencies that relate directly to the modeling process, such as understandings of the subjectivity of modeling processes, of the relationship between mathematics and reality, and of the perception of mathematics as a process rather than just a product. Some competencies are required for a successful and goal-oriented modeling process, such as the ability to communicate, the ability to work cooperatively in a group, and metacognitive competencies (Kaiser, 2007).

In the last two decades, a number of research projects have been carried out to foster modeling competencies (for an overview, see Niss & Blum, 2020). In summary, this research has found that modeling can be learned and that the development of modeling competencies can be supported by different measures. There are two ways to assess modeling competencies and their development in written form (Hankeln & Greefrath, 2020).

First, some instruments (e.g., Kreckler, 2015; Rellensmann et al., 2019; Schukajlow & Leiß, 2011) focus on modeling problems that have to be holistically worked on by students. The overall work is then evaluated using different coding schemes to obtain one value for a student's modeling competencies. Thus, modeling competencies are assessed in a holistic manner (Blomhoj & Jensen, 2003). In order to draw accurate conclusions about existing modeling competencies, it is advisable to have the students work on several of these problems (Klieme et al., 2008).

Second, test items are used for separate assessment of sub-competencies (e.g., Brand, 2014; Hankeln & Greefrath, 2020; Zöttl et al., 2011). Thus, modeling competencies are assessed in an atomistic manner (Blomhoj & Jensen, 2003). One disadvantage of a test designed in this way is that no statements can be made about whether an entire modeling process can be performed by students. The advantage, however, is that task-specific aspects are not very important because the sub-competencies are measured by several items. Furthermore, a test designed in this way allows statements to be made about sub-competencies that would otherwise not be measurable due to their dependence on previous steps in the modeling process. This particularly applies to the sub-competencies of interpreting, validating, and exposing, which cannot be applied without the sub-competencies of understanding, simplifying, and mathematizing.

In recent years, probabilistic test theory has been applied to evaluate measurement instruments with atomistically designed items. In addition, research has examined the extent to which the assumed subdivision can be confirmed empirically in the data. In previous studies, models with partly different dimensions best reflect the empirical data. Evaluations of tests with atomistic items have used both between-item models, in which the items load on exactly one underlying latent dimension, and within-item models, in which the items load on multiple latent dimensions, depending on which type of model better described the empirical data. In addition, the models differ in terms of accuracy for differentiating modeling sub-competencies. While the models of Brand (2014) and Zöttl et al. (2011) divided competencies into three latent dimensions – *from reality to mathematics, working mathematically*, and *from mathematics to reality* – the model of the LIMo Project (Hankeln & Greefrath, 2020) identified four sub-competencies: *simplifying, mathematizing, interpreting*, and *validating*. The results of these studies show that individual sub-competencies develop differently in different studies. On a methodological level, it can be said that in different studies, different models were best suited to describe the empirical data. It should be emphasized that the models differed in the extent to which modeling sub-competencies could be empirically separated.

Metacognitive strategies

Various scientific disciplines have researched the concept of metacognition and its use and impact on learning processes. As this research has been conducted with different approaches and interests, there is no uniform concept of metacognition. Instead, individual facets of metacognition are conceptualized differently across approaches. In recent years, research has focused on the importance of metacognition for group processes. Metacognition is usually divided into metacognitive knowledge and metacognitive strategies, often supplemented by an affective-motivational factor (e.g., Efklides, 2008; Veenman et al., 2006). Many conceptualizations include all elements of Weinert's (2001) definition of competence, which is why some publications speak of metacognitive competencies when referring to the use of metacognition in learning and work processes.

In this study, metacognitive strategies are focused on. With respect to these, there is a discussion about the extent to which unconscious and automatized use of metacognitive strategies may be considered metacognitive. According to Veenman (2011) and Hartman (2001), automation occurs after repeated use of strategies that were initially consciously employed. Thus, these authors evaluate these unconsciously used strategies as having a metacognitive origin. The authors distinguish between consciously applied metacognitive strategies and skills that are usually executed unconsciously but can become conscious when problems arise. If it is not possible to determine whether strategies are applied consciously or unconsciously, they are often referred to as metacognitive behavior.

The metacognitive strategies that are necessary for goal-oriented and effective modeling have been empirically reconstructed by authors such as Stillman and Galbraith (1998). This set of strategies includes the following

Strategies for planning work processes given the specific requirements of the modeling problem, the persons involved, and the general conditions, such as the time or materials available: This includes all activities that help to ensure that the modeling problem can be processed in a goal-oriented and collaborative manner. The established plan does not have to cover the entire solution process from beginning to end; it may initially contain only the first steps, and further steps may be planned later in the process.

Strategies for monitoring and, if necessary, regulating the work process: These concern both the personal level (e.g., monitoring the cooperation of all group members and, if necessary, requesting further work, as well as strategies for ensuring all participants' understanding, and, if necessary, providing explanations) and the content level (e.g., monitoring compliance with the established plan and, if necessary, returning to the old plan or creating a new one).

Strategies for evaluating and assessing the modeling process with the aim of optimizing the process next time: These strategies involve both the personal level (evaluation of collaboration and cooperation within the group) and the content level (evaluation of the work process, e.g., with regard to the usefulness of applied strategies). The strategies have both reflexive and prospective components. Vorhölter and Kaiser (2016) and Vorhölter (2018, 2019) illustrated how the described metacognitive aspects manifest in the processing of specific modeling problems.

Since modeling problems are usually worked on at least partially by small groups, it is relevant for research to consider not only the metacognitive competencies of individual group members but also social metacognition. Social metacognition depends on individual group members' metacognitive competencies and includes strategies aimed at planning, monitoring, regulating, and evaluating the group process. It is important to note that social metacognition cannot simply be presented as the sum of individuals' metacognition. For metacognitive competencies to be effective within group work, first, they must be introduced into the group modeling process (i.e., made available to the group), and second, they must be accepted by the other group members. Groups that work effectively benefit from both individual and group strategies (Siegel, 2012). For a group to effectively work together on a task, it is not only important that each individual group member understands the task but also that a common understanding exists within the group. In other words, students must not only plan and carry out their own activities but also must coordinate the activities of all group members and combine them in a meaningful way. In addition, they must monitor not only their own behavior but also that of others by, for example, asking questions, listening to others' ideas, and making their own suggestions (Goos & Geiger, 1995). This requires that students communicate with one another and coordinate their actions, explain their ideas to each other, and externalize their thoughts. Only by doing this can they share their knowledge and skills (e.g., Artzt & Armour-Thomas, 1992; Chalmers, 2009; Rogat & Adams-Wiggins, 2014).

The relationship between individual and social metacognition has yet to be researched at an empirical level. However, Efklides (2008) developed a theoretical model that integrates the different areas of metacognition as well as the individual and social levels, thus illustrating the relationships between individual metacognition and social metacognition and unconscious and conscious individual metacognition. One of the few theoretical conceptualizations of the role of metacognitive individual strategies in the modeling process was developed by Schukajlow (2011), who explicitly remarked that it is not possible to locate individual aspects of metacognitive strategies in specific phases of the modeling process. Rather, according to his description of cognitive and metacognitive processes for approaching modeling problems, metacognitive strategies are necessary at every stage of the modeling process. This is consistent with the conclusions of Stillman and Galbraith (2011), who found that metacognitive strategies were applied at all phases of the modeling process. However, neither of the aforementioned studies explicitly distinguishes between metacognitive individual and group strategies.

Different methods are used for measuring metacognitive strategies. A general distinction is made between offline methods, which perform a measurement before or after concrete situations (or, sometimes, independently of such situations), and online methods, which assess the use of metacognitive strategies during the working and learning process. Overviews have been presented by Veenman et al. (2006) and Veenman and van Cleef (2019). In general, previous research indicates that it is not yet possible to empirically confirm the theoretical distinction between metacognitive strategies that is presented above. For example, Rakoczy and Klieme (2005) report cross-loadings between the factors of metacognition, planning and structuring, and control and evaluation. In addition, Schraw and Dennison (1994) did not succeed in empirically confirming the assumed structure of eight scales (three subscales containing metacognitive knowledge and five subscales containing metacognitive strategies), only the differentiation of metacognitive knowledge and metacognitive strategies. These results suggest a strong correlation between metacognitive strategies.

Metacognitive strategies in modeling processes

The application of metacognitive strategies should not only enable effective and goal-oriented modeling processes but also help to identify and overcome difficulties (Stillman, 2004). Stillman (2011) differentiated between different metacognitive behaviors, observing that usage depends on students' experience and knowledge of metacognitive strategies. She links the success of a metacognitive behavior, based on the results of Goos (2002), to students' reactions to so-called red-flag situations in problem solving. These situations are regarded as triggers for metacognition, and subsequent actions can be classified as appropriate or inappropriate. One of these actions (metacognitive blindness) was further discussed by Ng (2010), who explored metacognitive behavior in groups and introduced the notion of partial metacognitive blindness. The term refers to a case wherein some members of the group perceive a problem but are unable to convince the other members that there is a need to respond.

Metacognitive behavior can be triggered situationally. For example, the perceived difficulty of the problem, which in the case of modeling problems includes the requirement to create a model oneself and to identify the aspects it requires, can influence the use of metacognitive strategies (Stillman, 2004; Weinert, 1984). Teacher interventions can lead to the application of metacognitive strategies, or the application of strategies can be triggered by such interventions (Stender, 2016). In addition, the perception of a red-flag situation (Goos, 2002; Stillman, 2011) can be considered a trigger of metacognitive behavior.

Metacognitive behavior and its triggers can be influenced by factors, such as (modeling-specific) preferences, and traits, such as one's attitude toward modeling (Maaß, 2007), motivation, and interest in general (e.g., Borkowski et al., 2000). Furthermore, a certain degree of strategy maturity (e.g., Artelt & Neuenhaus, 2010; Hasselhorn & Gold, 2006) is necessary, since the transfer of strategies to other situations and tasks cannot be taken for granted (Hartman, 2001; Niss & Blum, 2020; Veenman, 2011). Finally, domain-specific metacognitive knowledge influences the use of metacognitive strategies (e.g., Efklides, 2008), which in this case means that modeling-specific metacognitive knowledge (e.g., Frenken, 2021; Vorhölter et al., 2019) influences the use of metacognitive modeling strategies.

Few studies have empirically examined the relationship between the use of metacognitive strategies and students' modeling performance or development of modeling competencies. One such study is that of Schukajlow and Leiß (2011), who analyzed the relationship between the usage of metacognitive strategies and students' modeling performance. The results revealed no significant correlation between mathematical modeling competencies and students' self-reported metacognitive strategies. In a similarly designed study, Schukajlow and Krug (2013) analyzed the impact of asking students to develop multiple solutions on the use of strategies for planning and monitoring. Their findings indicated that this task had a greater influence on the number of strategies used by students than a teacher's request that they use these strategies. Later, the authors revealed that the number of developed solutions mediated the development of metacognitive strategies for planning and monitoring (Krug & Schukajlow, 2020). In this study, monitoring included strategies for monitoring of the work process as well as reflection during and after the work process. Thus, it also included strategies for evaluation. This was due to the fact that strategies for monitoring and evaluating could not be empirically separated based on the available data.

Hidayat et al. (2018) analyzed the connection between metacognition in mathematical modeling and modeling competencies. The authors used a questionnaire that examined four dimensions – awareness, cognitive strategy, planning, and self-checking – which they were able to empirically prove are distinct from one another. The four factors showed high correlation. It should be noted that, aside from "planning," the dimensions do not correspond to traditional differentiations of metacognitive strategies.

Empirical assessment of the influence of metacognitive strategies on modeling performance is still in its infancy, and it strongly depends on the conceptualization and measurement instruments that are used. Nevertheless, due to the numerous observations of modeling processes that have been documented in research, it is regularly assumed that metacognitive strategies not only are necessary for goal-oriented processing of modeling problems but also influence the development of modeling competencies (Blum, 2011).

Research questions and hypotheses

Summing up the theoretical considerations and empirical findings mentioned above, metacognitive strategies can be differentiated into individual and group strategies, both of which have an influence on modeling processes. However, the relationship between these strategies and the influence on modeling processes have not yet been verified. In addition, metacognitive strategies are assumed to be important for the development of modeling competencies. Thus, this paper addresses the following research questions.

What is the relationship between metacognitive individual and group strategies?

It is assumed that metacognitive individual strategies tend to be highly correlated with one another, as do metacognitive group strategies. No assumptions regarding the relationship between metacognitive individual and group strategies can be derived from existing theory.

How does the development of modeling sub-competencies and metacognitive individual and group strategies differ between interventions? Can the development of modeling sub-competencies be explained by metacognitive individual or group strategies?

It is assumed that the usage of metacognitive modeling strategies will increase significantly if their use is explicitly addressed by teachers and students are encouraged to use them. The same holds true for modeling sub-competencies; it is assumed that the use of these sub-competencies will increase significantly if they are enhanced through work on modeling problems and are explicitly reflected on afterward. Furthermore, it is assumed that metacognitive individual and group strategies positively influence the development of modeling sub-competencies.

Methods and instruments

Design of the study

The study design (see Figure 1) is based on a pre-post intervention design with two groups. In one group (hereafter referred to as the *me-group*), the metacognitive strategies that were used and deemed useful were explicitly discussed in class following the processing of each task. In the other group (hereafter referred to as the *ma-group*), the mathematical methods that were used during work on the modeling problem were intensively discussed following processing of the task. The project's intervention phase lasted from October 2016 to July 2017, with approximately one modeling problem processed per month between November 2016 and May 2017. Classes were assigned in pairs (using grade level and school social index as criteria) to the two intervention groups. Classes from the same school were assigned to the same intervention group because it was assumed that, due to the long intervention period (i.e., about one

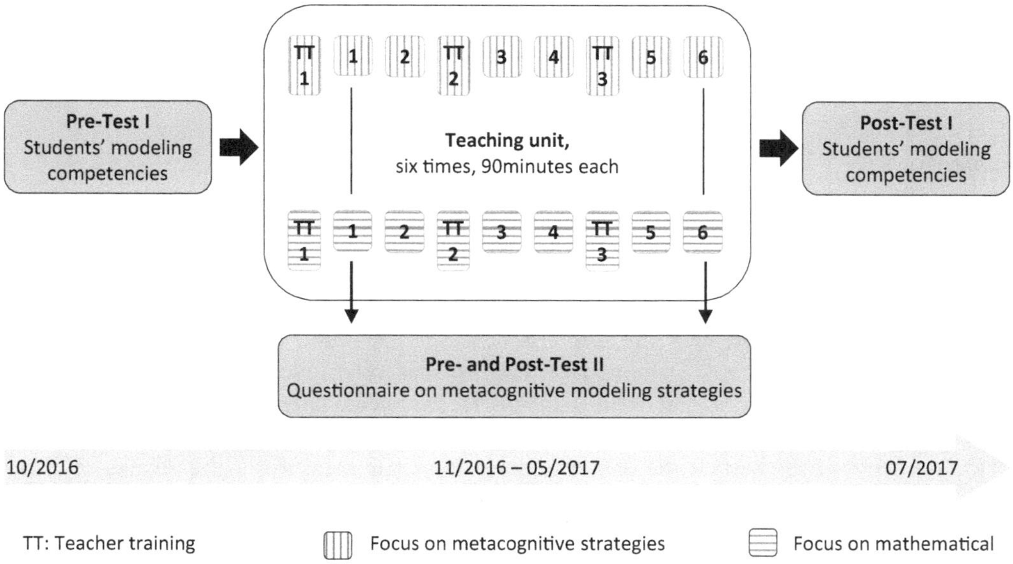

Figure 1. Design of the study with two intervention groups.

school year), both students within a grade and teachers within the school would exchange information about the project. Thus, separation of the two intervention groups could not be ensured. Modeling problems were worked on during mathematics lessons supervised by the students' math teacher. The researchers observed random lessons to ensure that the instructions and specified procedures were followed.

Before students worked on the first modeling problem, their modeling competencies were assessed using a test instrument adapted from the work of Brand (2014). Immediately after completing the first and sixth modeling problems, the students were asked to complete a questionnaire designed to collect self-reported information on the metacognitive modeling strategies they used. Approximately six weeks after the last modeling problem had been worked on, students' modeling competencies were measured a second time.

The participating teachers took part in three group-specific teacher training sessions before working on the first, third, and fifth modeling problems. At each teacher training session, the next two modeling problems were introduced, possible difficulties were discussed, and strategies for supporting the students were taught in accordance with the principles of minimal help and adaptivity. To support students, the teachers were asked to refer to the modeling process and use strategic aids. While teachers in the me-group discussed the advantages of using metacognitive strategies and ways to promote them in class, teachers in the ma-group discussed how they could deepen students' understanding of the mathematical methods they used when working on the modeling problems.

The implementation of the study was approved by the Institute for Educational Monitoring and Quality Development (IfBQ) in Hamburg.

Sample

Of the 532 students in grades nine and ten (thus aged between 14 and 16) who participated in the study, only 170 participated in all four measurement sessions. All available data sets were used to determine the factor scores, the individual ability parameters of the metacognitive strategies, and the latent dimensions of modeling sub-competencies. However, only the scores of students who participated in all four measurement sessions were used for the further analyses described in this paper. Of the 170 students, 77 (57% male, 43% female) were in the me-group and 93 (49% male, 51% female) were in the ma-group.

Instruments

In accordance with the study design, students' modeling sub-competencies and self-assessment of the metacognitive strategies they used were each measured at two time points.

Modeling test

To measure modeling competencies, the test items developed by Brand (2014) were used and, in some cases, slightly modified. In total, 61 items were used to measure modeling sub-competencies. More than one-third and less than one-half of the items per reconstructed latent dimension served as anchor items. Data were analyzed with the help of item response theory, comparing four different models. To select the model with the best fit the data, three psychometric measures – the Akaike information criterion (AIC), Bayes information criterion (BIC), and consistent AIC (CAIC) – were used. After model selection, weighted likelihood estimates (WLE) and individual ability parameters were determined.

Considering the psychometrical measures AIC, BIC, and CAIC, a three-dimensional between-item model provided the best description of the data. In addition, the three dimensions' EAP/PV reliabilities were acceptable, varying between .713 and .753. The first latent dimension includes *simplifying and mathematizing* (s/m) and thus covers the steps from understanding the real problem to development of the mathematical model. The second latent dimension, *working mathematically* (wm), includes the step from mathematical model to mathematical solution and concerns only inner-mathematical problems. The third latent dimension, *interpreting and validating* (i/v), comprises the activities from the mathematical world into reality.

Metacognitive self-assessment

A questionnaire using a 5-step Likert scale was developed to assess the perceived use of metacognitive strategies. The questionnaire covered both metacognitive individual and group strategies and aimed at strategy use before, during, and after working on a modeling problem. Immediately after the students claimed to have completed the problem, they were given the questionnaire individually. Each student rated their use of metacognitive individual and group strategies. Due to the clustered nature of the ratings on the use of group strategies, individual and group strategies were evaluated separately.

Analysis of individual strategies involved several steps. The goal was to reconstruct an invariant measurement model across groups and measurement times to ensure that the same construct was measured at all times and in all groups and to compare the scores between the groups (e.g., Steenkamp & Baumgartner, 1998).

First, metacognitive individual items were critically reviewed. Items that, due to their formulation, were very close to the validating step in the modeling process and items that showed low discriminatory power were excluded. With the remaining 13 items, an exploratory factor analysis (EFA) with promax rotation was performed on the data set of the me-group from the second measurement time. It was assumed that the concept of metacognition was only developed because of the intervention and, in the worst case, no invariant factor structure could be found in any group. In the course of the EFA, three items were excluded from further analyses due to statistically relevant cross-loadings (>.3). To determine the number of factors, a scree plot was used and a parallel analysis was performed.

In the second step, the fit measures and measurement invariance between subsamples was tested according to a common procedure (e.g., Hirschfeld & Brachel, 2014). In addition to changes in χ^2 and degrees of freedom (df), changes in the comparative fit index (CFI) and root mean square error of approximation (RMSEA) as well as the standardized root mean square residual (SRMR) within and between values were used to compare models (Chen, 2007; Cheung & Rensvold, 2002). Full scalar measurement invariance could not be confirmed, but partial scalar measurement invariance was determined by dissolving the restrictions regarding factor loadings for one item and the restriction for the intercepts of the same item as well as a second one. As the factors consist of more than two invariant items, students' factor scores can be compared with each other (Byrne et al., 1989; Steenkamp & Baumgartner, 1998).

Analysis of metacognitive individual strategies resulted in the reconstruction of three factors.[2] The first factor (*individual strategies for planning*, or IP) includes strategies for individual planning of the working process. The second factor (*individual strategies for controlling*, or IC) includes strategies for individual control of the working process and the result. The third factor (*individual strategies for regulation*, or IR) includes strategies for individual regulation in case of difficulties.

The method applied in this study to evaluate group strategies, which were measured independently from individual strategies, and individuals' self-assessments is described in detail by Vorhölter (2019). Therefore, it is presented here in general terms. The approach proposed by Hox et al. (2017) was used to analyze the data with a multi-level model. The analysis showed that the model, which has the same factor structure at the individual and group levels, had the best fit values when the time of measurement and intervention group were used as covariates ($\gamma 2 = 131$, df = 66, RMSEA = .04, $SRMR_{within} = .04$, $SRMR_{between} = .09$, CFI = .96). Constraining the factor loadings did not lead to significantly worse values. Thus, it can be assumed that there is measurement invariance across levels.

Analysis of the metacognitive group strategies revealed three factors. Strategies used to understand the task, plan, and monitor were subsumed into one factor, termed *group strategies for proceeding* (GP). Strategies used to regulate behavior when difficulties occur were termed *group strategies for regulating* (GR). Strategies for evaluating the entire process were termed *group strategies for evaluating* (GE).

Data analysis

To address the research questions and test the hypotheses, analyses were performed on the scores of the factors of metacognitive individual and group strategies as well as the individual ability parameters of the three latent dimensions of modeling competencies. After selecting the students to be considered, the values of each of the six factors and the three latent dimensions were scaled with a mean of 0 and a standard deviation of 1. The scores can be found in Table 1. All analyses were performed using SPSS (version 27).

Pearson correlations were calculated to determine the relationships between metacognitive individual and group strategies at the factor level. As standardized scores were used, outliers were not expected.

To assess the development of the three factors of metacognitive individual strategies, the three factors of metacognitive group strategies, and the three reconstructed latent dimensions of modeling sub-competencies, a repeated-measure multivariate analysis of variance (MANOVA) was performed

Table 1. Mean and standard deviation (SD) in the two groups and at both measurement times (MT I and MT II) .

Ma-group					Me-group			
	Mean		SD		Mean		SD	
	MT I	MT II	MT I	MT II	MT I	MP II	MT I	MT II
IP	0.04	−0.07	1,12	0.98	−0.01	0.04	0.95	0.93
IC	0.11	−0.21	1.02	1.04	0.17	−0.04	0.95	0.94
IR	−0.01	−0.06	1.08	0.93	−0.03	0.11	1.00	0.99
GP	−0.09	−0.03	0.84	1.14	0.00	0.13	0.94	1.05
GR	0.00	−0.05	0.90	1.06	0.03	0.03	1.02	1.03
GE	−0.09	−0.04	1.02	1.05	−0.11	0.25	0.85	1.02
s/m	−0.12	0.01	0.96	1.00	0.09	0.04	0.95	1.09
wm	−0.24	0.28	0.87	1.10	−0.19	0.14	0.82	1.09
i/v	−0.04	0.17	1.24	0.68	−0.02	−0.14	1.05	0.93

with group as a between-factor. The reason for this was the partially high correlation of the factors and latent dimensions, through which alpha error accumulation could have occurred if analyses of variance were separately performed. If the result of the MANOVA indicated significant differences in individual factors or latent dimensions, corresponding t-tests were performed and the effect size was determined.

Finally, moderator analysis was used to determine whether developments in the latent dimensions of modeling sub-competencies were influenced by metacognitive individual or group strategies. The analyses were performed using the PROCESS macro (version 3.5.3; Hayes, 2018) in SPSS.

Results

The following section presents the results of the analyses in regard to the two research questions.

Relationship between metacognitive individual and group strategies

Correlation analyses were performed to determine the relationships between the factors of metacognitive individual and group strategies. The values of the correlations can be found in Table 2.

At the beginning of the study, there were high positive correlations between all metacognitive individual strategies. The same was true at the second measurement time. With regard to the metacognitive group strategies, there were significant positive correlations in both groups and at both measurement times (with the exception of GE and GR, which were correlated in the ma-group at the first measurement time).

There were low to medium correlations between metacognitive individual and group strategies at both times and in both groups. The correlations were not significant in the me-group, except for the correlation between GP and IR. In the ma-group, the correlations between GE and metacognitive individual strategies were not significant at either measurement time. However, at the first measurement time, the correlations between the other two metacognitive group strategies GP and GR and metacognitive individual strategies were significant. At the second measurement time, there were no longer significant correlations for GP and IP or IR.

Development of modeling sub-competencies and metacognitive strategies

To address the second research question, which concerns the development of modeling sub-competencies and metacognitive modeling strategies, the scores of the latent dimension of modeling sub-competencies and the factors of the metacognitive individual and group strategies were used in a multivariate repeated-measures MANOVA in which the intervention was applied as a within-subject factor. Within the two groups, there were significant developments in four dimensions/factors

Table 2. Correlations between metacognitive strategies at the first measurement time (left side) and the second measurement time (right side). Above the diagonal are the values of the me-group (highlighted in red), and under the diagonal are the values of the ma-group (highlighted in blue). *p < .05; **p < .01, ***p < .001.

	First measurement time						Second measurement time					
	IP	IC	IR	GP	GR	IP	GE	IC	IR	GP	GR	GE
IP		.71***	.82***	.33**	.31**	.15		.71***	.87***	.19	.09	−.01
IC	.85***		.63 ***	.27*	.19	.23*	.82***		.57***	.11	.19	.16
IR	.87***	.71***		.20	.14	.12	.89***	.80***		.27*	.18	.00
GP	.33**	.31**	.32**		.83***	.35**	.20	.21*	.19		.76***	.53***
GR	.34***	.36***	.32**	.70***		.29*	.23*	.25 *	.26*	.80***		.32**
GE	.04	.07	.02	.34***	.17		.17	.13	.18	.64***	.43***	

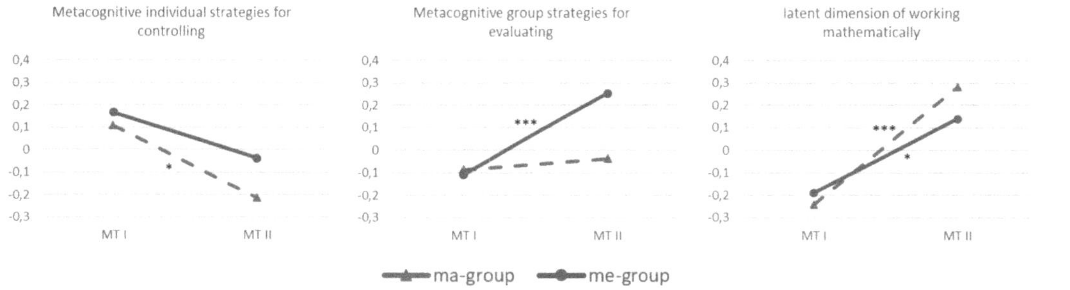

Figure 2. Developments in metacognitive individual strategies for controlling (left side), metacognitive group strategies for evaluating (in the middle), and the latent dimension of working mathematically (right side), differentiated for the ma-group (blue line with triangles) and me-group (red line with circles). $*p < .05$; $**p < .01$, $***p < .001$.

The MANOVA showed a significant change in IC ($F(1, 5.88) = 8.87$, $p < .001$, *partial* $\eta^2 = .05$). As shown in Figure 2, students reported that they used these strategies less often at the second measurement time. A one-tailed t-test of the dependent samples revealed a small effect size for the ma-group (t $(92) = 2.71$, $p = .04$, $d = .28$) and a non-significant trend in the me-group ($t(76) = 1.56$, $p = .06$).

Regarding the development of GE, the results of the total sample of 532 students (Vorhölter, 2019) were confirmed for the sample of 170 students used in this analysis (see Figure 2). In other words, analysis showed a significant increase in the use of these strategies ($F(1, 3.53) = 4.37$, $p = .04$, *partial* $\eta^2 = .03$). A closer look at the development using one-tailed t-tests for dependent samples in the two groups showed that the increase within the me-group was significant with a small effect size ($t(76) = 2.68$, $p < .001$, $d = .31$), but the development in the ma-group was not significant ($t(92) = .35$, $p = .36$)

A significant change in the latent dimension of wm was found over time ($F(1, 15.3 = 33.27)$, $p < .01$, *partial* $\eta^2 = .17$). Analysis of the one-tailed t-test for both groups showed that, unlike for GE, the effect size in the ma-group was higher ($t(92) = 5.23$, $p < .001$, $d = .54$) than in the me-group ($t(76) = 3.03$, $p = .01$, $d = .35$). Both effects were at a medium level (see Figure 2). Moderation analysis was also performed to determine whether the development of this latent dimension was moderated by metacognitive strategies. The results did not show significant effects for any of the factors of metacognitive individual or group strategies.

With respect to the latent dimension of i/v, neither of the two main effects (i.e., intervention and time) showed any significant influence. However, the interaction of these two effects did ($F(1, 2.24 = 4.14$, $p = .04$, *partial* $\eta^2 = .02$). The development in both groups (see Figure 3) suggested that no significant change occurred over time, nor did a significant change occur in each group individually. However, the intervention resulted in an increase in the latent dimension of i/v in the ma-group and a decrease in the me-group. Thus, the difference between the groups at the second measurement time was significant, as confirmed by a one-tailed t-test for the independent samples (t $(168) = 2.5$, $p = .01$).

Moderation analyses were conducted to determine whether metacognitive individual and group strategies had an influence on the development of i/v. The overall model was significant ($F(3, 166) = 18.12$, $p = .04$), predicting 24.67% of the variance. Moderation analysis showed that IC was the only factor that moderated the development of the latent dimension of i/v ($\Delta R^2 = 2\%$, $F(1, 166) = 2.31$, $p = .04$). However, the bootstrap results suggested that this moderation is not significant because the confidence interval included zero (95% $CI[-.005, .234]$). Thus, moderation analyses were conducted separately for both groups.

In the ma-group, the overall model was significant ($F(3, 89) = 4.77$, $p < .01$), predicting 13.86% of the variance. However, the moderation was not significant. In the me-group, the overall model was significant ($F(3, 73) = 20.31$, $p = .01$), predicting 45.5% of the variance. Moderation analysis showed that, in this group, IC moderated the development of the latent dimension of i/v ($\Delta R^2 = 5.13\%$, $F(1, 73) = 6.88$, $p = .01$, 95% $CI[.027, .43]$).

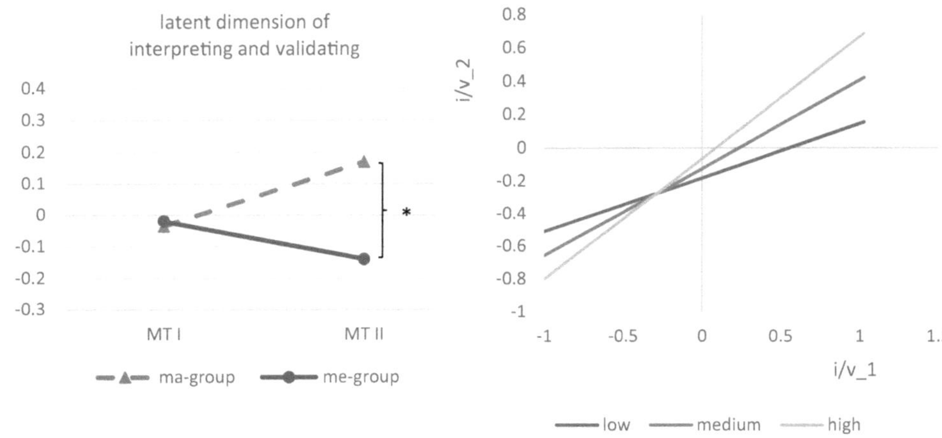

Figure 3. Left side: Developments of the latent dimension of interpreting and validating, differentiated for the ma-group (blue line with triangles) and me-group (red line with circles). Right side: Visualization of the moderation analysis of the development of this latent dimension within the me-group at the first (i/v_1) and second (i/v_2) measurement times, differentiated for students with low (- 1 SD) and medium and high (+ 1 SD) factor scores for metacognitive individual strategies for controlling.

Discussion

Individual metacognitive strategies were highly and significantly correlated, both at the beginning and end of the study and in both groups. Two of the three metacognitive group strategies – GP and GR – were also highly correlated (at the second measurement point) in both groups. GE had low to medium correlation at the first measurement time, but medium to high correlation at the second measurement time with GP and GR. The hypothesis that the correlation between strategies is higher in the group of students who are familiar with the concept of metacognition (me-group) could not be confirmed (Niss & Blum, 2020). However, the results indicated that metacognitive individual strategies are highly correlated with each other and that metacognitive group strategies are highly correlated with each other (with one exception). This aligns with the results of previous studies, including those of Schraw and Dennison (1994), who could not differentiate between sub-scales of metacognitive strategies, and those of Rakoczy and Klieme (2005), who reported cross-loadings between different facets of metacognitive strategies.

The low correlations between metacognitive individual and group strategies are striking. One reason for the low correlation may be that factors of individual and group strategies, with the exception of the strategies for regulation, target different areas. While individual strategies are empirically divided into planning and control of the work process across both measurement times, these strategies remain – as demonstrated by Vorhölter (2018), – grouped together as GP. Differentiation may be attributed to the fact that the teachers in both groups discussed the importance of individual work phases with the students, particularly at the beginning of group work phases. The metacognitive strategies that become important during the process of modeling are supplemented by strategies for regulation for both individual students and groups. Consequently, a separate area of planning can only be empirically proven to exist at an individual level. Unlike at the first measurement time (see Vorhölter, 2018), strategies for evaluating can only be reconstructed empirically as a range of group strategies in the data set. It can be assumed that this is an effect of the learning environment. In the me-group, the importance of group work – and of evaluating the group's working processes – was intensively discussed with the students, both at the person and task levels. This is evidenced by the significant increase in this area. General behavioral patterns during group work phases were addressed in the ma-group.

The weaker correlation between metacognitive individual and group strategies at the second measurement time (compared to the first measurement time) could be explained by the fact that, because the students were working on complex problems in groups on several occasions, the students became more proficient in these strategies and were more able to differentiate between individual and group behavior.

The low correlation between metacognitive individual and group strategies underlines the assumption of Siegel (2012): groups benefit from both metacognitive individual and group strategies, since these are different facets of metacognition.

Concerning the development of metacognitive modeling strategies, significant developments were found only for IC and GE. IC decreased significantly in both groups, and GE increased in both groups, but this was significant only within the me-group.

There may be several reasons for the decrease in IC. First, the students indicated that they had used these strategies very frequently at the first measurement time. It is possible that they learned to more critically evaluate their strategies for controlling during the intervention. Second, it is possible that, during the intervention, the small groups assigned individual students to use these strategies for the entire group, meaning that the other students used them less.

GE increased significantly according to the students' self-assessment, but only in the me-group. It should be noted that this area was the first to be promoted in the intervention group and was repeatedly discussed with the students throughout the project. Therefore, it is possible that the use of these strategies reached the level of strategy maturity (e.g., Artelt & Neuenhaus, 2010; Hasselhorn & Gold, 2006).

When evaluating the usage of metacognitive individual and group strategies, it is important to consider the approach to measuring them. For example, the usage of single strategies was rated by self-assessment, and some strategies had high measures at the beginning of the study. It may be that students' self-assessments changed during the study, so the same measures at different times during the study may have different meanings. Furthermore, as mentioned above, some measures of areas of metacognitive strategies, particularly those for individual and group regulation, can only be evaluated meaningfully when it is known whether problems occurred.

Regarding working mathematically as a latent dimension of modeling sub-competencies, there was an increase in both groups, with a greater effect in the ma-group. This indicates that the explicit thematization and deepening of previously known mathematical procedures led to stronger support for these procedures. The processing of modeling problems in itself also supported this latent dimension. The mathematical procedures used by the students when working on the modeling task – and discussed by the teachers – are basic, and may include calculating volumes, calculating within a right-angled triangle, or handling variables and equations. The scores of students in the me-group, who applied these procedures while processing the modeling problem but did not discuss them in class after working on the modeling problem significantly increased in the latent dimension of working mathematically. This indicates that the processing of modeling problems within this study promoted not only metacognitive strategies but also basic mathematical skills.

The latent variable of i/v, as part of modeling sub-competencies, decreased significantly in the me-group, moderated by IC. The results of the moderation analysis indicate that students with a high factor score for IC were able to show competencies in relation to interpreting and validating at the second measurement time. In contrast, those with low scores for IC recorded a drop in competencies. There is no obvious reason why students' scores decreased. The extent to which this relationship is not only statistical but also substantive cannot be meaningfully stated here, since it is the only reconstructed influence of metacognitive modeling strategies on the development of modeling sub-competencies.

In summary, regarding the development of the latent dimension of modeling sub-competencies, metacognitive individual and group strategies did not have a significant influence on the increase of latent dimension of modeling sub-competencies.

Implications for future research

At first glance, the results presented in this paper seem to be discouraging. While the three factors of metacognitive individual strategies and the three factors of metacognitive group strategies correlate with one another, almost no significant correlation was found between metacognitive individual and group strategies at the end of the study. Hardly any significant increase in the use of metacognitive individual and group strategies or modeling sub-competencies was revealed. Metacognitive individual and group strategies showed a significant influence on the development of modeling sub-competencies in only one case.

However, taking into account additional information given by participating teachers and students within the framework of the overall project (which was not used in this analysis), the results have important implications for future research.

Analysis of the development of the use of metacognitive modeling strategies did not reveal any significant increases, except for GE. In contrast, for the same sample, Krüger (2021) reported that students' use of metacognitive strategies became increasingly autonomous over the course of the study.

From a methodological point of view, it should be questioned whether the use of metacognitive modeling strategies can be measured by students' self-reports. As stated previously, the students may have reported high values for certain strategies at the beginning of the study. Thus, it must be questioned whether the self-assessments at the beginning and end of the study were made on the same basis.

Furthermore, it can be assumed that the students' understanding of the construct of metacognition became more differentiated over the course of the study, especially in the me-group. This assumption is supported by the less significant correlation between individual and group strategies at the second measurement time. To ensure consistency in students' rating of their use of metacognition, it may be possible to familiarize the students with the construct of metacognition and the differences between group and individual strategies before beginning a study. This recommendation is supported by a case-based analysis of existing self-reports in combination with data in the form of interviews with the students and video recordings of the work processes. It is possible that the information in self-reports could be better understood when the setting is considered (Vorhölter & Krüger, 2021). Another possibility is to observe students' behavior, and have experts assess it. However, such a procedure could exclude non-verbalized metacognitive individual strategies and would be very cost- and time-intensive for large samples.

With regard to modeling competencies, only relatively small changes in the reconstructed latent dimensions could be detected. These slight changes are not surprising in light of other studies (e.g., Brand, 2014; Hankeln & Greefrath, 2020). Although this is a sobering result, it emphasizes the importance of adequate measurement instruments for assessing modeling competencies. This study evaluated only sub-competencies, not comprehensive modeling competencies or the ability to independently work on holistic modeling problems (within a group).

This leads to the second implication concerning the measurement of modeling competencies. As stated by several participating students and their teachers, it became easier for the students to work on the modeling problems over the course of the study. This confirms the complexity of modeling competencies as a construct. Furthermore, it is clear that measuring modeling sub-competencies is not sufficient if the relationship between metacognition and modeling competencies must be considered. The use of metacognitive strategies is considered to be effective for either preventing problems from arising in the first place or identifying them as quickly as possible (strategies for planning and controlling); solving problems as effectively as possible (strategies for regulation); or improving the process in the long term (strategies for evaluation; see also Goos, 2002; Stillman, 2011). This means that the involved metacognitive strategies affect the entire process, not simply individual sub-steps.

Before approaching revision of the instruments, data from the overall project should be triangulated. The aforementioned case-based study by Vorhölter and Krüger (2021) should be extended to a larger sample, and the assessment of metacognitive strategies should be adjusted on this basis. Furthermore, the results of Wendt (2021) concerning the teachers who participated in the study should be used to analyze differences in the development of modeling sub-competencies as well as metacognitive strategies in the different classes and intervention groups. This could allow for more precise statements regarding the influence of certain characteristics of the classes. Currently, the ability for such statements is limited due to the non-randomized (but pairwise) assignment of the classes to the intervention groups.

In addition to the methodological implications for future research, it should be emphasized that, according to the participating students and teachers, the students' modeling competencies increased. Empirically, it was found that GE could be promoted in both groups, but the increase in these strategies was only significant in the me-group. It was also found that the latent dimension of working mathematically, which referred to already-known procedures, increased in both groups during the study, although to a greater extent in the ma-group.

These results mean that both interventions are suitable for promoting metacognitive strategies and mathematical skills. Thus, in addition to the acquisition of modeling competencies, working on modeling problems can result in the acquisition of mathematical skills.

Notes

1. Sometimes, *working mathematically* is not regarded as a sub-competence of mathematical modeling, as it is not considered to be specific to mathematical modeling.
2. The factors reconstructed in the analyses presented in this paper differ from the three factors mentioned by Vorhölter (2018). At a methodological level, the different factors can be attributed to different samples and the claim of measurement invariance. In terms of content, one can assume that, by working on modeling problems several times in small groups, students independently developed a construct of metacognition.

Disclosure statement

No potential conflict of interest was reported by the author(s).

ORCID

Katrin Vorhölter http://orcid.org/0000-0003-3141-2916

References

Artelt, C., & Neuenhaus, N. (2010). Metakognition und Leistung. In W. Bos (Ed.), *Schulische lerngelegenheiten und kompetenzentwicklung: Festschrift für Jürgen Baumert* (pp. 127–146). Waxmann.

Artzt, A. F., & Armour-Thomas, E. (1992). Development of a cognitive-metacognitive framework for protocol analysis of mathematical problem solving in small groups. *Cognition and Instruction*, *9*(2), 137–175. https://doi.org/10.1207/s1532690xci0902_3

Blomhoj, M., & Jensen, T. H. (2003). Developing mathematical modelling competence: Conceptual clarification and educational planning. *Teaching Mathematics and Its Applications*, *22*(3), 123–139. https://doi.org/10.1093/teamat/22.3.123

Blum, W. (2011). Can modelling be taught and learnt? Some answers from empirical research. In G. Kaiser, W. Blum, R. Borromeo Ferri, & G. A. Stillman (Eds.), *International perspectives on the teaching and learning of mathematical modelling. Trends in teaching and learning of mathematical modelling* (Vol. 1, pp. 15–30). Springer. https://doi.org/10.1007/978-94-007-0910-2_3

Blum, W. (2015). Quality teaching of mathematical modelling: What do we know, what can we do? In S. J. Cho (Ed.), *The proceedings of the 12th international congress on mathematical education* (pp. 73–96). Springer International Publishing. https://doi.org/10.1007/978-3-319-12688-3_9

Borkowski, J. G., Chan, L. K., & Muthukrishna, N. (2000). 1. A process-oriented model of metacognition: Links between motivation and executive functioning. *Issues in the Measurement of Metacognition*, (2), 1–41 https://digitalcommons.unl.edu/burosmetacognition/2.

Borromeo Ferri, R. (2018). *Learning How to teach mathematical modeling in school and teacher education*. Springer International Publishing. https://doi.org/10.1007/978-3-319-68072-9

Brand, S. (2014). *Erwerb von Modellierungskompetenzen: Empirischer Vergleich eines holistischen und eines atomistischen Ansatzes zur Förderung von Modellierungskompetenzen (Aufl. 2014)*. Springer Fachmedien Wiesbaden.

Byrne, B. M., Shavelson, R. J., & Muthén, B. (1989). Testing for the equivalence of factor covariance and mean structures: The issue of partial measurement invariance. *Psychological Bulletin*, *105*(3), 356–466. https://doi.org/10.1037/0033-2909.105.3.456

Chalmers, C. (2009). Group metacognition during mathematical problem solving. In R. K. Hunter, B. A. Bicknell, & T. A. Burgess (Eds.), *Crossing divides: Merga 32 conference proceedings, 5-9 July 2009, massey university, Wellington, New Zealand* (pp. 105–111). MERGA.

Chen, F. F. (2007). Sensitivity of goodness of fit indexes to lack of measurement invariance. *Structural Equation Modeling: A Multidisciplinary Journal*, *14*(3), 464–504. https://doi.org/10.1080/10705510701301834

Cheung, G. W., & Rensvold, R. B. (2002). Evaluating goodness-of-fit indexes for testing measurement invariance. *Structural Equation Modeling: A Multidisciplinary Journal*, *9*(2), 233–255. https://doi.org/10.1207/S15328007SEM0902_5

Efklides, A. (2008). Metacognition. *European Psychologist*, *13*(4), 277–287. https://doi.org/10.1027/1016-9040.13.4.277

Frenken, L. (2021). Measuring students' metacognitive knowledge of mathematical modelling. In F. K. S. Leung, G. A. Stillman, G. Kaiser, & K. L. Wong (Eds.), *Mathematical modelling education in east and west* (pp. 215–225). Springer. https://doi.org/10.1007/978-3-030-66996-6_18

Goos, M., & Geiger, V. (1995). Metacognitive activity and collaborative interactions in the mathematics classroom: A case study. In W. F. Atweh & S. Flavel (Eds.), *Proceedings of the 18th conference of the mathematics education research group of australasia* (pp. 307–313). MERGA.

Goos, M. (2002). Understanding metacognitive failure. *Journal of Mathematical Behavior*, *21*(21), 283–302. https://doi.org/10.1016/S0732-3123(02)00130-X

Greefrath, G., & Vorhölter, K. (2016). *Teaching and learning mathematical modelling: Approaches and developments from german speaking countries. ICME-13 topical surveys*. Springer. https://doi.org/10.1007/978-3-319-45004-9

Hankeln, C., & Greefrath, G. (2020). Mathematische Modellierungskompetenz fördern durch Lösungsplan oder dynamische Geometrie-Software? Empirische Ergebnisse aus dem LIMo-Projekt. *Journal Für Mathematik-Didaktik*, 367–394 . https://doi.org/10.1007/s13138-020-00178-9

Hartman, H. J. (2001). Developing students' metacognitive knowledge and skills. In H. J. Hartman (Ed.), *Neuropsychology and cognition: Vol. 19. Metacognition in learning and instruction: Theory, research and practice* (2nd ed., 33–68). Kluwer.

Hasselhorn, M., & Gold, A. (2006). *Pädagogische psychologie: Erfolgreiches Lernen und Lehren* (1st ed.). *Kohlhammer Standards Psychologie*. Kohlhammer.

Hayes, A. F. (2018). *Introduction to mediation, moderation, and conditional process analysis: A regression-based approach* (2nd ed.). *Methodology in the social sciences*. The Guilford Press.

Hidayat, R., Syed Zamri, S. N. A., & Zulnaidi, H. (2018). Does mastery of goal components mediate the relationship between metacognition and mathematical modelling competency? *Educational Sciences: Theory & Practice*, *18*(3): 579–604.

Hirschfeld, G., & Brachel, R. (2014). Multiple-group confirmatory factor analysis in r – A tutorial in measurement invariance with continuous and ordinal indicators. *Practical Assessment, Research & Evaluation*, *19*(7), 1–12. https://doi.org/10.7275/qazy-2946

Hox, J. J., Moerbeek, M., & van de Schoot, R. (2017). *Multilevel analysis: Techniques and applications, third edition* (3rd ed.). *Quantitative Methodology Series*. Taylor and Francis.

Kaiser, G., & Brand, S. (2015). Modelling competencies: Past development and further perspectives. In G. A. Stillman, W. Blum, & M. Salett Biembengut (Eds.), *Mathematical modelling in education research and practice* (pp. 129–149). Springer. https://doi.org/10.1007/978-3-319-18272-8_10

Kaiser, G. (2007). Modelling and modelling competencies in school. In C. Haines, P. L. Galbraith, W. Blum, & S. Khan (Eds.), *Mathematical modelling (ICTMA): Education, engineering and economic* (pp. 110–119). Horwood Publishing.

Kaiser, G. (2017). The teaching and learning of mathematical modelling. In J. Cai (Ed.), *Compendium for research in mathematics education* (pp. 267–291). National Council of Teachers of Mathematics.

Klieme, E., Hartig, J., & Rauch, D. (2008). The concept of competence in educational contexts. In J. Hartig, E. Klieme, & D. Leutner (Eds.), *Assessment of competencies in educational contexts* (pp. 3–22). Hogrefe.

Kreckler, J. (2015). *Standortplanung und Geometrie*. Springer Fachmedien Wiesbaden.

Krug, A., & Schukajlow, S. (2020). Entwicklung prozeduraler Metakognition und des selbstregulierten Lernens durch den Einsatz multipler Lösungen zu Modellierungsaufgaben. *Journal Für Mathematik-Didaktik*, *41*(2), 423–458. https://doi.org/10.1007/s13138-019-00154-y

Krüger, A. (2021). *Metakognition beim mathematischen Modellieren: Strategieeinsatz aus Schülerperspektive*. Springer.

Maaß, K. (2007). Modelling in class: What do we students want to learn? In C. Haines, P. L. Galbraith, W. Blum, & S. Khan (Eds.), *Mathematical modelling (ICMTA12): Education, engineering and economic* (pp. 63–78). Horwood Publishing.

Ng, K. E. D. (2010). Partial metacognitive blindness in collaborative problem solving. In L. Sparrow, B. Kissane, & C. Hurst (Eds.), *Shaping the future of mathematics education: Proceedings of the 33rd annual conference of the mathematics education research group of australasia* (pp. 446–453). MERGA.

Niss, M., & Blum, W. (2020). *The learning and teaching of mathematical modelling. Impact: Interweaving mathematical pedagogy and content for teaching*. Routledge.

Rakoczy, K., & Klieme, E. (2005). *Dokumentation der Erhebungs- und Auswertungsinstrumente zur schweizerisch-deutschen videostudie. „Unterrichtsqualität, Lernverhalten und mathematisches Verständnis".: 1. Befragungsinstrumente. Materialien zur Bildungsforschung: Vol. 13*. GFPF [u.a.].

Rellensmann, J., Schukajlow, S., & Leopold, C. (2019). Measuring and investigating strategic knowledge about drawing to solve geometry modelling problems. *ZDM Mathematics Education*, *52*(1), 97–110. https://doi.org/10.1007/s11858-019-01085-1

Rogat, T. K., & Adams-Wiggins, K. R. (2014). Other-regulation in collaborative groups: Implications for regulation quality. *Instructional Science*, *42*(6), 879–904. https://doi.org/10.1007/s11251-014-9322-9

Schraw, G., & Dennison, R. S. (1994). Assessing metacognitive awareness. *Contemporary Educational Psychology*, *19*(4), 460–475. https://doi.org/10.1006/ceps.1994.1033

Schukajlow, S., & Krug, A. (2013). Planning, monitorin and multiple solutions while solving modelling problems. In A. M. Lindmeier & A. Heinze (Eds.), *Mathematics learning across the life span: Proceedings of the 37th conference of the international group for the psychology of mathematics education* (pp. 177–184). PME.

Schukajlow, S., & Leiß, D. (2011). Selbstberichtete Strategienutzung und mathematische Modellierungskompetenz. *Journal Für Mathematik-Didaktik*, *32*(1), 53–77. https://doi.org/10.1007/s13138-010-0023-x

Schukajlow, S. (2011). *Mathematisches Modellieren: Schwierigkeiten und Strategien von Lernenden als Bausteine einer lernprozessorientierten Didaktik der neuen Aufgabenkultur. Empirische Studien zur Didaktik der Mathematik*. Waxmann.

Siegel, M. A. (2012). Filling in the distance between US: Group metacognition during problem solving in a secondary education course. *Journal of Science Education and Technology*, *21*(3), 325–341. https://doi.org/10.1007/s10956-011-9326-z

Steenkamp, J. -. B. E. M., & Baumgartner, H. (1998). Assessing measurement invariance in cross-national consumer research. *Journal of Consumer Research*, *25*(1), 78–107. https://doi.org/10.1086/209528

Stender, P. (2016). *Wirkungsvolle Lehrerinterventionsformen bei komplexen Modellierungsaufgaben. Perspektiven der Mathematikdidaktik*. Springer.

Stillman, G. A., & Galbraith, P. L. (1998). Applying mathematics with realworld connections: Metacognitive characteristics of secondary students. *Educational Studies in Mathematics*, *36*(2), 157–194. https://doi.org/10.1023/A:1003246329257

Stillman, G. A., & Galbraith, P. (2011). Evolution of applications and modelling in a senior secondary curriculum. In G. Kaiser, W. Blum, R. Borromeo Ferri, & G. A. Stillman (Eds.), *International perspectives on the teaching and learning of mathematical modelling. Trends in teaching and learning of mathematical modelling* (Vol. 1, 689–699). Springer. https://doi.org/10.1007/978-94-007-0910-2_66

Stillman, G. A. (2004). Strategies employed by upper secondary students for overcoming or exploiting conditions affecting accessibility of applications tasks. *Mathematics Education Research Journal*, *16*(1), 41–71. https://doi.org/10.1007/BF03217390

Stillman, G. A. (2011). Applying metacognitive knowledge and strategies in applications and modelling tasks at secondary school. In G. Kaiser, W. Blum, R. Borromeo Ferri, & G. A. Stillman (Eds.), *International perspectives on the teaching and learning of mathematical modelling. Trends in teaching and learning of mathematical modelling* (pp. 165–180). Springer.

Veenman, M. V. J., Hout-Wolters, B., & Afflerbach, P. (2006). Metacognition and learning: Conceptual and methodological considerations. *Metacognition and Learning*, *1*(1), 3–14. https://doi.org/10.1007/s11409-006-6893-0

Veenman, M. V. J., & van Cleef, D. (2019). Measuring metacognitive skills for mathematics: Students' self-reports versus on-line assessment methods. *ZDM Mathematics Education*, *51*(4), 691–701. https://doi.org/10.1007/s11858-018-1006-5

Veenman, M. V. J. (2011). Learning to self-monitor and self-regulate. In P. A. Alexander & R. E. Mayer (Eds.), *Educational psychology handbook series. Handbook of research on learning and instruction* (pp. 197–218). Routledge.

Vorhölter, K., & Kaiser, G. (2016). Theoretical and pedagogical considerations in promoting students' metacognitive modeling competencies. In C. R. Hirsch & A. R. McDuffie (Eds.), *Annual perspectives in mathematics education: Annual perspectives in mathematics education 2016: Mathematical modeling and modeling mathematics* (pp. 273–280). National Council of Teachers of Mathematics.

Vorhölter, K., Krüger, A., & Wendt, L. (2019). Metacognition in mathematical modeling – An overview. In S. A. Chamberlin & B. Sriraman (Eds.), *Advances in mathematics education. Affect in mathematical modeling* (pp. 29–51). Springer.

Vorhölter, K., & Krüger, A. (2021). Metacognitive strategies in modeling: Comparison of the results achieved with the help of different methods. *Quadrante, 30*(1), 178–197. https://doi.org/10.48489/QUADRANTE.23653

Vorhölter, K. (2018). Conceptualization and measuring of metacognitive modelling competencies: Empirical verification of theoretical assumptions. *ZDM Mathematics Education, 50*(1–2), 343–354. https://doi.org/10.1007/s11858-017-0909-x

Vorhölter, K. (2019). Enhancing metacognitive group strategies for modelling. *ZDM Mathematics Education, 51*(4), 703–716. https://doi.org/10.1007/s11858-019-01055-7

Weinert, F. E. (2001). *Leistungsmessungen in Schulen. Beltz-Pädagogik.* Beltz.

Weinert, F. E. (1984). Metakognition und motivation als determinanten der Lerneffekitvität: Einführung und Überblick. In F. E. Weinert, R. H. Kluwe, & A. L. Brown (Eds.), *Metakognition, motivation und Lernen* (pp. 9–21). Kohlhammer.

Wendt, L. (2021). *Reflexionsfähigkeit von Lehrkräften über metakognitive Schülerprozessen beim mathematischen Modellieren* [Dissertation]. Universität Hamburg.

Zöttl, L., Ufer, S., & Reiss, K. (2011). Assessing modelling competencies using a multidimensional irt approach. In G. Kaiser, W. Blum, R. Borromeo Ferri, & G. A. Stillman (Eds.), *International perspectives on the teaching and learning of mathematical modelling. Trends in teaching and learning of mathematical modelling* (Vol. 1, pp. 427–437). Springer. https://doi.org/10.1007/978-94-007-0910-2_42

Mathematical Thinking about Systems – Students Modelling a Biometrics Identity Verification System

Ana Margarida Baioa and Susana Carreira

ABSTRACT

The aim of this study is to understand how students' mathematical thinking is activated and nurtured in solving a modeling problem, where the problem situation involves the design of a system. From a STEM integrated perspective, 9th grade students worked on a modeling task aiming to create an identification system based on hand biometrics. The theoretical framework proposes a conceptualization of the interplay between the mathematical modeling process, from a cognitive perspective, and the engineering design process. Central ideas refer to the cyclical nature of both processes and to the sub-processes involved in them. The empirical data were collected in two design-based research cycles with different 9th grade classes. The data from the groups' audio and video recording and the students' productions were analyzed under a directed qualitative content analysis informed by theory. The results showed a global pattern in the students' thinking in solving a design system problem. The overlapping and interplay between the mathematical modeling and the design process was a prominent characteristic of students' thinking. The modeling cycle was mirrored by a design cycle, with both running in parallel. System thinking pushed and drove students' mathematical thinking, from the system requirements to the prototype validation.

Introduction

Our contemporary world poses increasingly complex problems for citizens and decision-makers at the regional, national, and global level. Due to their complexity, tackling such problems requires the articulation and integration between different areas of knowledge and expertise. This is particularly the case of problems that are linked to the design and development of entire systems, which are pervasive and essential in human activity, such as transport systems, surveillance systems, security systems, electrical systems, and many others (Aslaksen, 2013). Put in a simple way, a system is composed of a set of elements that are related to each other, through internal relations, and also related to the outside environment, through external relations (Hubalovsky, 2010). Most real-world systems are highly intricate and raise problems where concepts, results and procedures from mathematics, science, technology, and engineering are not only required but must be effectively connected.

In recent years, a concern with more interdisciplinary approaches within the school curricula, the inclusion of project-based activities and inquiry-based learning addressing authentic open real-world problems has become more evident. This advocacy of interdisciplinary STEM blends with the trend for promoting critical and creative thinking, problem solving, and digital literacy, in education (OECD, 2018). It therefore drives educational movements toward hands-on activities, STEM practices, design thinking, interdisciplinary problem solving, and engineering projects (English, 2019; Leung, 2019; Tytler et al., 2019).

Several researchers have developed and proposed theoretical views on the interplay between mathematical modeling and interdisciplinary education (e.g., Borromeo Ferri & Mousoulides, 2017; English, 2009; Michelsen, 2006, 2015) and likewise between mathematical modeling and STEM education (e.g., Araya Schulz, 2016; Hallström & Schönborn, 2019; Kertil & Gurel, 2016).

Despite these efforts, the role and place of mathematical modeling regarding the interconnection between mathematics and other subjects is not apparent and clearly defined, whether from a cognitive perspective or from a didactical perspective.

For example, Borromeo Ferri and Mousoulides (2017) note that interdisciplinarity is not immediately visible in the modeling cycle that underlies much of the research and practice on teaching and learning mathematical modeling. Thus, in many modeling problems, it is the modeler who, at some point, finds the need for extra-mathematical knowledge and understands that such knowledge is linked to other disciplines or scientific domains. From this point of view, interdisciplinarity becomes very close to mathematical modeling, although not every interdisciplinarity task is necessarily a modeling task. Moreover, when such overlapping takes place, – in what could be called interdisciplinary modeling activities – the mathematical modeling process acquires a multi-dimensional nature: "Speaking on an abstract level, the individual modelling routes are on a multi-dimensional level, when the other discipline(s) included in the modelling problems is understood by the students" (Borromeo Ferri & Mousoulides, 2017, p. 902). In fact, the students must work with both mathematical and scientific concepts, which requires crossing boundaries between domains of knowledge. Such boundary crossing, as already mentioned, is not portrayed in the well-known modeling cycle, and even a form of representing such an interplay seems to remain a challenge.

Another similar claim is made by Leung (2019), when discussing the relationship between mathematical modeling and an integrated STEM pedagogical framework for mathematics and science. As in the previous case, the idea of a boundary pedagogy is supported under the following perspective: "a *boundary pedagogy* that can *cross the boundaries* of the STEM disciplines and serves as a boundary object to connect the classroom practices of the disciplines (Leung, 2019, p. 1344, original emphasis). The author speaks of a concoction that brings together mathematical modeling, inquiry-based learning, and tool-based pedagogical environments. The proposal includes a tentative inquiry-based modeling cycle that intends to bridge scientific inquiry and mathematical modeling. It is described as a "hybrid boundary object attempting to merge inquiry-based learning and mathematical modeling with an intentional ambiguity between them which allows the opening of discourse space and hence translation between the two pedagogies" (Leung, 2019, p. 1345). The suggested cycle does not actually reveal the extra-mathematical knowledge or the interdisciplinary work. Furthermore, in that cycle, the elements pertaining to the inquiry-based approach consist of sub-processes that are very close to the sub-processes found in various modeling cycles from a cognitive perspective (e.g., Blomhøj & Jensen, 2003; Blum & Leiβ, 2007; Haines & Crouch, 2010). Therefore, it is reasonable to claim, as Leung concludes, that mathematical modeling, inquiry-based learning, and engineering design involve several common processes and that all those approaches have problem-solving at its core. However, much remains to be known about how the different approaches intersect or may be effectively combined in the educational context.

Research context and aim

Our research work has been progressively developed within the scope of a doctoral research project, in which the first author takes the role of a teacher-researcher in a middle school (students aged 13–15 years). Following a design-based research methodology, we implemented two design cycles centered on solving mathematical modeling tasks in the classroom, whose pedagogical objectives are consistent with an integrated STEM approach.

The main characteristics of such a task are summarized as follows: i) the task presents a problem intending to replicate a real-world situation (e.g., sometimes a response to a customer's request is wanted); ii) the problem to be solved involves the design and analysis of a system (or part of a system) that exists in the real world; iii) the solution to the problem entails hands-on experimentation with physical objects and instruments, some being of daily use; iv) answering the problem includes achieving a (simplified) prototype of the real system, for which the construction of a mathematical model is indispensable.

Drawing on the data gathered so far, we first concentrated our attention on aspects of task design. We have focused on studying the ways in which middle school students reacted to the use of experiments and materials as part of the mathematical modeling taking place in the class. Therefore, we considered the role of experimentation and use of physical materials as distinctive elements in mathematical modeling activities that involve the design of a system (Carreira & Baioa, 2018). More recently, we have reexamined our empirical data by mainly concentrating on the cognitive processes developed by the students. Specifically, from a semiotic point of view, we have looked at how the experimental work carried out by the students contributes to the construction of meaning for the models produced (Carreira et al., 2020).

Moving forward the work presented earlier (Baioa & Carreira, 2019a), mainly at the Eleventh Congress of the European Society for Research in Mathematics Education - CERME 11, we are pursuing our research in seeking to better understand how the students' modeling, in particular the mathematical thinking involved in the modeling, is related to the process of designing a real system or part of a system, which includes a hands-on approach, involving experimentation, and building prototypes.

The present study addresses the data collected in 9th grade classes, taking place in different school years, in which the students worked on a task proposing the construction of a hand biometrics recognition system that could be potentially used in the school to grant access to authorized persons only (for instance, in a computer lab). The following research question reflects our interest in achieving a sharper understanding of the types of relationships, connections, and mutual influences that can be identified and may constitute pedagogical benefits, between the modeling process and the engineering design process, particularly significant in an integrated STEM context. We thus formulate our research question:

How is mathematical thinking activated and nurtured in solving a modeling problem, when the problem is linked to the design of a system or part of a system?

This research aim, in bringing mathematical thinking to the forefront, is relevant to the well-established study field of mathematical modeling as, to the present, a theory is missing that clearly highlights the role of mathematical modeling and of its sub-processes in the development of STEM activities, from an interdisciplinary perspective. There is still much to be known and investigated about interdisciplinary work, where mathematical modeling plays a central role (Tezer, 2020).

Theories about teaching and learning in STEM education are not as effectively structured as the solid theoretical conceptualization of mathematical modeling in education. Nevertheless, there is much scattered evidence that mathematical modeling and the construction of mathematical models are an essential activity for understanding real-world systems, as well as for the development of prototypes, and this can be observed with students across a wide range of age levels (Araya Schulz, 2016; Baioa & Carreira, 2019a; Bock et al., 2019; English, 2019; English et al., 2017).

Theoretical background

Our theoretical background brings forward a mapping of research perspectives and theoretical developments leading to a conceptualization of the interplay between mathematical thinking and real-world system design situations where mathematical modeling is a central cognitive process.

Mathematical modeling from a cognitive perspective

Schukajlow et al. (2018) claim that the empirical research on teaching and learning mathematical modeling has predominantly assumed a cognitive focus over the last 40 years. The emphasis on the cognitive perspective has largely resulted from an increasing presence of mathematical modeling problems, tasks, and projects in various curricula around the world. This motivated the need to understand how students and teachers perform the mathematical modeling process and, subsequently, to identify and find strategies to overcome potential blockages. At the same time, interest has grown in proposing idealized schematic forms that allow describing, with reasonable pragmatism, how a real-world problem is transformed into a mathematical problem and solved mathematically to generate results that must be interpreted in terms of the situation, thus leading to the revision or validation of the model.

The modeling cycle from a cognitive perspective

Several authors in the field of mathematical modeling in education have long assumed that the modeling process has a cyclical nature, in addition to being interpreted as a sequence of steps with the transitions corresponding to the completion of specific or atomized sub-processes. Haines and Crouch (2010) have argued that several possibilities of visually representing the so-called modeling cycle are unequivocally useful as a means for understanding the ideal behavior of a modeler moving from a real-world problem to a mathematical model and to an acceptable solution. Nevertheless, they stressed that the modeling process "in the classroom and amongst modellers, in industry and elsewhere, is not like that and is likely to be much more complex" (p. 148). In fact, there is a reasonably consensual understanding that the modeling cycle is an ideal representation of a complex process and that its cyclical nature is far from straightforward and linear. Instead, different types of analyzes of the students' cognitive processes show that there is a strong idiosyncratic nature in their paths. Substantial evidence can be found, for example, in studies focusing on recreating individual modeling routes (e.g., Blum & Borromeo Ferri, 2009; Borromeo Ferri, 2006, Borromeo Ferri, 2007a, 2007b, 2018; Ramírez-Montes et al., 2021) or on producing descriptive chronological diagrams of the modeling activity (e.g., Albarracín et al., 2019; Ärlebäck, 2009; Czocher, 2016).

Although this continues to be, in many ways, a topic of debate, there are still no signs that the modeling cycle is about to be replaced, either in terms of theoretical conceptualization or as an analytical tool, by other constructs that are at least equally productive. Indeed, English et al. (2016) identified three prevalent theoretical approaches to the mathematical modeling process, apart from the various perspectives related to the goals of applications and modeling in mathematics teaching and learning adopted in different studies (Kaiser & Sriraman, 2006). The characterization of the modeling process based on a cycle that connects the mathematical domain and the rest of the world is predominantly featured, often embracing the modeling cycle proposed by Blum and Leiβ (2007), as presented in Figure 1.

It should be noted that most of the studies aiming to understand the students' cognitive processes that take place in solving modeling problems somehow adhere to the modeling cycle, either as a theoretical construct that captures the process or as an analytical tool that supports the reading and interpretation of empirical data. Borromeo Ferri (2007b) condenses the cognitive perspective by stating that "under a cognitive perspective the focus lies on the individual thinking processes which are expressed mainly through certain verbal and non-verbal actions in combination with written solutions during modeling activities of individuals (including teachers)" (p. 265). Equally important is the statement made by Kaiser and Sriraman (2006) that a cognitive perspective "starts from a descriptive position" (p. 307). It refers to the analysis of the various modeling processes that can be observed and documented when individuals or groups solve modeling problems of different types and with different levels of complexity.

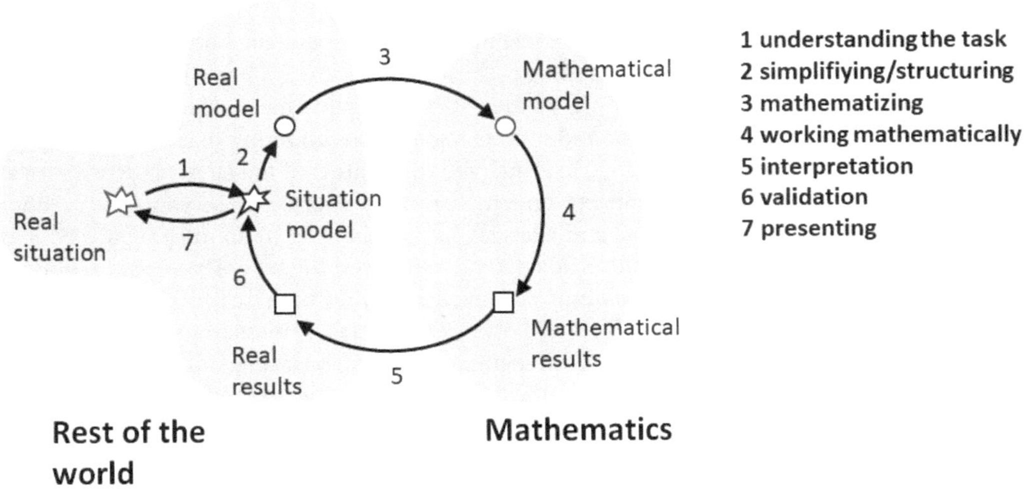

Figure 1. Visual representation of the mathematical modeling cycle (Blum & Leiβ, 2007).

The interplay between mathematical and extra-mathematical knowledge

Among many others, three studies that comply with a cognitive standpoint are relevant to our theoretical background, as they highlight the non-linear and largely ill-defined character of students' modeling processes. Furthermore, they clearly set the crucial role that the modeler's knowledge bases play in the modeling activity.

The first study, by Czocher (2016), uses an analytical tool that combines the modeling cycle, including its sub-processes, with modeling activity diagrams (MAD). The analysis focuses on the transitions performed by the modeler and is operationalized through modeling transition diagrams (MTDs), which expose individual modeling routes and give a chronological account of the modeling sub-processes. The results obtained with engineering undergraduate students solving different modeling tasks showed the absence of universal patterns in the obtained diagrams and suggested that an ideal MTD may not even exist. The research findings also pointed to the importance of the interaction between the modeler's knowledge bases, which led to the conclusion: "students' progress in mathematical modeling is highly dependent on their own mathematical thinking and nonmathematical knowledge" (p. 98).

The second study was reported by Almeida (2018) and aimed to know how students use mathematics when solving modeling problems. Focusing on the processes of mathematization by undergraduates pursuing a degree in mathematics, the author examined the students' ability to anticipate the mathematics to be used, drawing on a theoretical discussion of the concept of mathematization proposed by Niss (2010). According to the results, the process of mathematization is connected to the use of a mathematical apparatus; however, that apparatus must be seen by the students as being useful and effective to generate a mathematical model and to provide an answer to the idealized situation. This means that the mathematical apparatus depends on the student's knowledge base, but it also hinges on the individual's anticipation of the productive connection between such apparatus and some idealized model of the situation. As asserted by Niss (2010), "successful mathematization requires *implemented anticipation* of a rather involved nature" (p. 56, original emphasis).

Finally, we consider the research study by Brown and Edwards (2011). Their work draws on the ideas proposed by Stillman (2000), on the relevance of prior knowledge of the modeling situation or context, and on the ideas of Busse (2005) regarding the relationship that the solver can establish between the reality and the mathematics. Concerning the dominance of extra-mathematical knowledge or mathematical knowledge in the solver's approach to a modeling task, four typical cases can be identified: reality bound, mathematics bound, integrating, and ambivalent. The authors present and discuss in detail a case of an integrating approach by students in upper secondary school. They relate this integrating fashion of dealing with the modeling tasks to the students' deepening their understanding of mathematics and their understanding of the modeling process. Brown and Edwards (2011) suggest that an integrative approach between mathematics and reality means the ability to synthesize mathematics and reality in solving a task and reflects higher order thinking that includes reasoning, communication, and metacognitive skills. They analyzed the context of the tasks proposed – relating to environmental issues and their impacts on economy and species preservation – as offering opportunities for engaging with the goal and allowing to respond to it in an integrated way. Similar results were obtained by Carreira and Baioa (2011) with middle school students working on a modeling task that involved manipulating and experimenting with real objects (folding cake boxes). The study showed that the hands-on activity, including trials, measuring, and actual folding and unfolding, offered students the chance of deepening their understanding of the real situation and of the mathematics involved. In this case, the authors concluded that deeper conceptual understanding was enabled by the process of experimentation embedded in the task.

Summarizing the most salient ideas that arise from these studies undertaking a cognitive perspective of mathematical modeling, it is worth mentioning that the modeling cycle, emphasizing the various sub-processes involved, is a prevalent theoretical tool. Its ideal character contrasts with empirical results that show a non-linear process manifested by several transitions back and forth between different steps. Nonlinearity is closely related to the interplay between the modeler's knowledge bases. The integration between mathematical knowledge and non-mathematical knowledge seems to be closely associated with the way the task context offers opportunities to activate and synthesize these types of knowledge allowing a deep understanding of the situation and of the mathematical model produced. Hence, there is significant interest in better understanding how mathematical thinking is put into practice when students are faced with a real situation involving the design of a system or a part of a system.

System design in STEM education

Activities focused on STEM education typically encourage students to make connections between real-world contexts and different forms of knowledge, including everyday, scientific, mathematical, and technological knowledge. New knowledge is expected to be produced around big ideas, concepts, or themes, in such a way as to create links between concepts and procedures, situated in a specific context or real situation. STEM activities are also concerned with the development or construction of a system, or a part of a system, and the resulting knowledge is fundamentally related to a discourse and practice that is socially produced over time. As several authors have described it, systems thinking is an approach to thinking holistically about the world that seeks a comprehensive understanding of real-world problems and phenomena, by using and combining multiple and diverse perspectives (Salado et al., 2019).

STEM education undergoes a currently evolving effort to integrate science, technology, engineering, and mathematics, based on authentic problems and tasks, taking a holistic approach to real-world problems, and promoting the integration of knowledge and practices from several fields (Breiner et al., 2012; English, 2016, 2015; Kelley & Knowles, 2016; Kertil & Gurel, 2016; Michelsen, 2006, 2015; Moore & Smith, 2014). Many of the studies reporting on cases of integrated STEM education involve teaching of an experimental nature, namely using lab experiments, simulations, prototyping, or field work, and the learning context often seeks to put the scientific process or the engineering design process in action (Stohlmann et al., 2012). As pointed out by Kelley and Knowles (2016), "science education can be enhanced by infusing an engineering design approach because it creates opportunity to apply science knowledge and inquiry as well as provides an authentic context for learning mathematical reasoning for informed decisions during the design process" (p. 5).

Some specific studies have focused on combining engineering and mathematics and, in several cases, the mathematical modeling cycle and the modeling process are used as the conceptual framework for organizing and guiding the work to be developed with the students (e.g., Araya Schulz, 2016; Diefes-Dux et al., 2008; Gallegos, 2018; Kertil & Gurel, 2016; Mousoulides & English, 2008, 2011). Exploring the many sides of a real system or engaging in the design of new products and processes are among the aims of the tasks proposed in those studies. The reference to a hypothetical client to whom an answer or a prototype is to be provided seems to be a relevant element in this type of task, since it contributes to making the context more credible for the student, and to deliver a solution not as an abstract result but as something to be employed in the real world by a presumed receiver (Carreira & Baioa, 2018).

The design process from a STEM perspective

English (2019) presents a strong case for the integration of STEM through design. One of the main claims is the fact that design problem-solving enables very relevant learning during the entire process that the design entails. Like many other researchers, she describes the design process as a cyclical process composed of several phases: (a) problem framing; (b) generation of ideas for planning and construction; (c) designing and constructing; (d) testing and assessing outcomes; (e) redesigning and reconstructing; (f) reflecting and communicating. Although the author did not focus specifically on the mathematical modeling process and rather sought to understand distributed learning across the various areas of STEM, the findings of her study, in which elementary school children were challenged to design their own pair of shoes, showed significant learning across all areas, including mathematics (measurement, statistics, and spatial reasoning). Finally, the results revealed that the students actively explored the roles of designers and engineers in shoe manufacture.

The study developed by Kertil and Gurel (2016) involved preservice teachers and denotes the intention to relate the process of mathematical modeling with design projects, as in the case of the project proposed in their study that consisted of building a rocket model. In their work, the design process is also assumed as a cyclical and iterative process. Another relevant aspect of their theoretical conceptualization is the fact that they consider physical models (the rocket models) and analytical models (the scientific and mathematical models) involved in the construction of efficient rockets. The authors underline the coinciding goals of mathematical modeling in education and integrated STEM education, stating that mathematical modeling can provide a bridge to STEM: "to some extent, mathematical modeling as a process is involved in all STEM related applications" (p. 46).

A finer analysis of the role of models and modeling in STEM activities is presented in the article by Hallström and Schönborn (2019). Here the discussion of the nature and function of models is thoroughly detailed, based on the review of fifteen key STEM education contributions. The authors bring a multifaceted model conception, based on its capacity to simplify reality: "models simplify aspects of reality and range from simple conceptual sketches and rough prototypes to advanced mathematical models" (p. 3). They argue that modeling involves a set of skills related to STEM education, which include: understanding what a model is and how to use it; carefully defining the context of the modeling process; mentally visualizing the model outcome; defining the mode of representation to express the model in. Another idea that stands out is the importance of students recognizing and understanding the functional/intentional nature of a model, especially when it is incorporated in design problems. Thus, they draw attention to the need to take both the user's and the designer's point of view into account in authentic design situations.

Key features of design and prototyping

In the field of education, there are still very few research studies that have focused on the explicit role of models and of the modeling process in engineering design problems. As such, it is reasonable to look outside the field to gain insights into the design and prototyping process and the modeling that takes place in that context.

The affinities between the modeling process and the prototyping process are quite evident, to the point that it is difficult to establish a clear boundary that allows distinguishing a prototype from a model. One of the key ideas of the engineering design process is that engineers need to somehow model the processes and products they want to create (Aslaksen, 2013; Dym, 2004; Dym et al., 2005). Models, especially mathematical models, are essential for predicting how a system or process will behave before the actual engineering product is finished. Other than mathematical, physical models are also greatly used in the design process. A physical model provides a structure that produces behaviors, reproduces some aspects of a system, displays a variety of reactions to the individual's actions, and works as a means of generating hypotheses while exploring a problem. Physical models are, therefore, objects and mechanisms that can be used in various ways to create knowledge, in the sense that they materialize theories, ideas and concepts (Knuutila, 2005).

Fatima and collaborators, in an article on the rapid application development process (Fatima et al., 2018), state that prototyping is the most efficient methodology to build a model of an intended system. For defining the system requirements, all the information and problems deemed relevant are gathered. This helps to understand the system. The prototype is nothing more than a limited working model of the imagined system. Essentially, prototyping is the process that leads to the simplified realization of a concept or idea about a product to be developed that reproduces essential aspects of the behavior of the final product. The prototyping process has a cyclical character and can be schematically represented by means of the prototyping cycle (Figure 2).

Isa and Liem (2014) explain that mathematical modeling and prototyping are closely associated. While the first aims to generate a mathematical model of some part of reality, the second aims to study and test how a new product will be used and how it will work in its production state. In the various engineering domains, prototypes integrate physical models, mathematical models, and computational models. All of them are indispensable representation tools, allowing to express ideas, concepts, features, properties, constraints, etc., as accurately and effectively as possible. They are also necessary to carry out testing, simulation, or experimentation. According to Birta and Arbez (2007), "the modeling activity creates an object (i.e., a model) that is subsequently used as a vehicle for

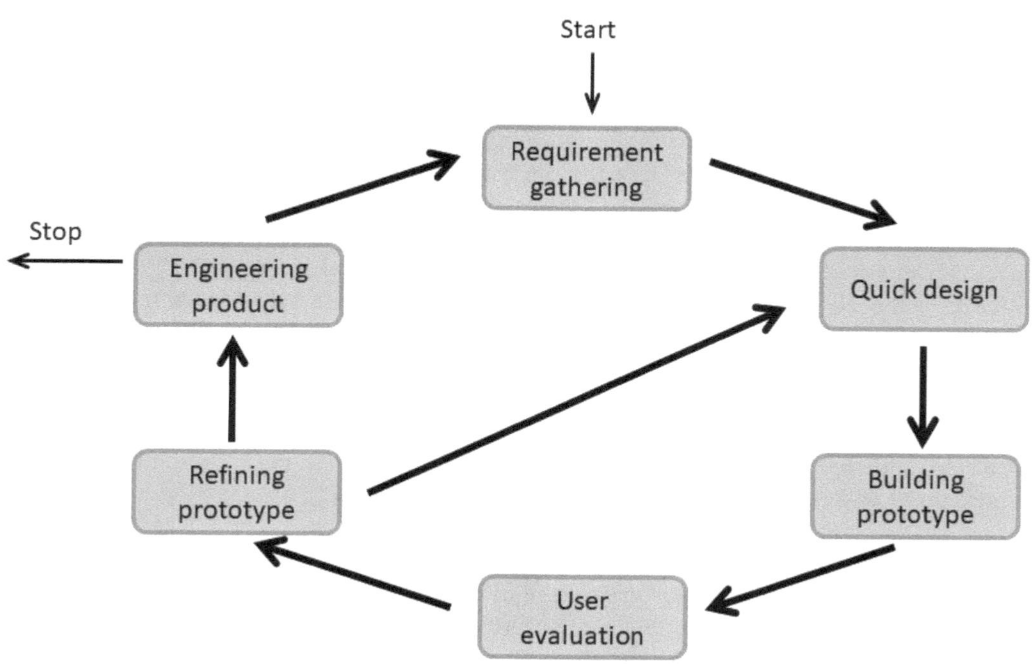

Figure 2. The prototyping cycle (adapted from Laudon & Laudon, 2014).

experimentation. This experimentation with the model is the simulation activity" (p. 4). Experimentation or simulation reinforce the cyclic character of the engineering design; the process may restart (entirely or in part), depending on the results of testing the prototype and validating the models.

In the present study we take the design process in structuring tasks that aim at creating simple prototypes of real-world systems in the class. Such tasks typically present the need to collect and use real data, and to connect knowledge and concepts from various STEM disciplines. Another important feature of such design problems is the construction of a mathematical model as an integral part of the system to be designed.

Method

In this section the methodological procedures of the present study are specified. The participants and the adopted methodology, which involved two cycles of design-based research, are presented. We describe in detail the task that is considered in this study, which proposes the design of a hand-biometrics recognition system (Appendix 1). We also define the qualitative data analysis approach and the coding scheme that was produced to examine the empirical data.

Participants

The study involved three classes of 9th graders (ages range 13–15 years-old) in two school years, corresponding to the first and second cycles of a design-based research (DBR): two classes were involved in the first cycle (DBR1) and a third class in the second cycle (DBR2). The participants were 46 students (27 male and 19 female) of two classes in the DBR1, and 18 students (5 male and 13 female) of one class in the DBR2, all from the same school located in a coastal city in Portugal.

The choice of a design-based research methodology was motivated by the developmental character of the study (Cobb, 2000; Kelly et al., 2008). Conducting a teaching experiment over two cycles of design, implementation, and evaluation was considered an adequate strategy to assess conjectures and possibilities regarding the students' modeling processes, but also to refine and improve tasks with specific characteristics for planning and testing strategies of STEM integration in which mathematical modeling is essential.

The hand-biometrics recognition task (Appendix 1) was the third task proposed. It was solved in a class period of 100 minutes. A subsequent class period of 50 minutes was used for collective discussion about the models and prototypes produced. The students organized themselves in work groups of 3–5 elements and received identical sets of materials for the activity. The teacher (first author) supervised the groups' work and interacted with the students, either to ask them for explaining their ideas or to give support and feedback (for instance, drawing attention to the requirements of the system) when some felt unable to move forward.

The empirical data were collected through audio and video recording of the groups, as well as photos documenting their work; the students' written reports were also collected.

Task and materials

During the research and development project, the students were exposed to several activities, each presenting a realistic and credible problem related to developing a real-world system. Examples of such problems were: modeling a tintometric system, used for customizing wall paint in a dispensing machine that discharges the correct amounts of colored pigments; modeling a fruit calibration system based on sliding fruits on a V-shaped rail; and modeling a hand-biometrics verification system to control individuals' access to a particular area of the school. All tasks comprised some form of practical work, carried out in the classroom, as a central part of creating a system or part of a system (see Baioa & Carreira, 2018, Baioa and Carreira, 2019b, for details on the tasks and teaching experiment).

The data reported in this article refer to the task concerning the creation of a hand-biometrics recognition system. The task was conceived to resemble the design process of verification systems based on hand geometry. The Encyclopedia of Biometrics (Li & Jain, 2009) provides concise and significant information on the foundations of hand-shaped recognition systems and their evolution over the years. One of the entries, by Duta (2009), presents the following definition (p. 682):

> A hand shape biometric system uses a camera or scanner-based device to acquire the hand image of a person from which shape information is extracted and compared against the information stored in a database to establish identity. Due to its limited discrimination power, a hand shape biometric system mostly operates in verification mode; that is the system *confirms* or *negates* the claimed identity of an individual.

According to researchers in hand biometrics and computer engineering (Duta, 2009; Varchol & Levický, 2007), the architecture of such a recognition system is composed of three stages: enrollment, processing, verification. The enrollment stage consists of generating a database of hand templates of registered users. Such templates contain a set of features extracted from the hand images collected for each user. Many features (palm width and height, finger widths and heights at various points, etc.) are assessed for their discrimination quality and the number of features is narrowed down. The processing stage involves matching, which means creating a method for comparing a user feature against the templates stored in the database and generating matching scores. This requires the development of a comparison algorithmic procedure, typically based on a distance calculation model and a threshold value for acceptance or rejection. In the verification stage, a decision is made based on the threshold defined; if the matching score is below a given threshold the claimed identity is accepted, otherwise it is rejected.

In the class, the students worked with a set of full-size printed images of their own hands (obtained by means of black and white photocopies). Before the class, each student made two photocopies of the palm of the hand, taking the hand up and putting it back on the glass top after the first photocopy. The set of first images was used by the groups to collect data; the set of second images was used to test the system. In addition, a printed outline of a human hand was given as a potential resource for data representation and recording.

To develop a prototype of the verification system, each group would use the data obtained from the images of the integrating members' hands. That system would operate in such a way that the members of the group would get authorized access and anyone outside the group would get denied access. For testing and validating the prototype, the students could use one or more randomly chosen images (belonging or not to the group) from the second set of photocopies.

Data analysis approach

The analysis of the qualitative data has a descriptive and interpretive nature and follows a qualitative content analysis method, more specifically a directed content analysis (Hsieh & Shannon, 2005; Potter & Levine-Donnerstein, 1999).

The transcripts of the audio and video and the written productions were reviewed and coded according to pre-defined categories (see Table 1 for codes, descriptors, and categories). The coding scheme was produced using the background theory to generate the codes and categories. Assuming a cognitive perspective (Borromeo Ferri, 2007b), the codes comprise students' non-verbal actions (what the students do) and students' verbal actions or ideas (what the students say and discuss). The deductive approach was adopted based on the research question, which seeks to know how

Table 1. Coding scheme.

Code	Code descriptor	Category	Theme
Examine images Explore images Discuss hand shape Discuss system purpose	- students compare shapes of hands ... - students draw points, lines, segments, circles, polygons ... - students discuss similarity, details of hands ... - students discuss security, reliability, accuracy ...	Understanding	*Hand Shape*
Use images Make measurements Use instruments Discuss hand features Discuss hand anatomy Discuss system purpose	- students consider geometrical elements (lengths, areas, perimeters ...) - students define measurements, decide units ... - students use ruler, drawing compass, set square ... - students define hand characteristics ... - students discuss hand parts, fingers, joints, phalanges, palms ... - students discuss the better way to describe a hand ...	Structuring	*Hand Features*
Organize data Represent data Discuss system operation	- students use labels, names, indexes ... - students create tables, lists, vectors ... - students discuss identification, database, authorized and unauthorized user ...	Mathematizing	*Data Collection*
Analyze data Build a model Discuss the model Discuss procedures Discuss concepts	- students compare values, consider variability, closeness ... - students perform calculations, define rules, use mathematical tools ... - students discuss comparison, discrimination, matching ... - students discuss algorithms, acceptance, rejection ... - students discuss error, approximation, distance ...	Working mathematically	*Decision Model*
Apply the model Obtain results Discuss results	- students use randomly chosen hands, collect features, use the model ... - students make calculations, perform mathematical procedures, achieve results - students obtain a decision, check a decision against the user identity ...	Interpreting	*Prototype Validation*
Evaluate the model Improve the model Perform trials Discuss flaws Discuss improvements	- students approve or disapprove the model ... - students change the model to obtain correct decisions ... - students repeat tests with different hands, adjust tolerance, repeat measurements ... - students discuss wrong results, make new assumptions, revise values ... - students discuss new results, evaluate the prototype ...	Validating	

mathematical thinking is activated and nurtured in solving a modeling problem related to the design of a hand-biometrics recognition system. Therefore, the theoretical framework on mathematical modeling and system design is at the basis of the qualitative data analysis.

Two rounds of data coding were independently carried out by two researchers (the authors). In the first round, a high rate of intercoder agreement was obtained (close to 93%). The few divergences were resolved with new independent and new joint coding attempts, and this led to a revision of some of the codes and their descriptors. After the first coding round, the codes to be used and their descriptors were stabilized. The data analysis was conducted for each of the work groups, using the coding scheme.

The second round of encoding was more interpretive, seeking to detect a general pattern in the data across the groups. A global pattern was obtained, composed of five themes that allow description and characterization of the thinking processes developed by the various groups of students. The five themes were added to the codes and categories (Table 1, 4th column) and used as a support structure to produce results and their discussion.

The anonymity of the participants was ensured, with the groups named with a capital letter and students identified by the respective group letter and a number.

Results and discussion

The data analysis unveiled a common pattern in the students' activity and thinking on the task. This global pattern is presented and described next, with the most significant variations between the various groups being pointed out.

Short transcripts of the students' exchanges, from different groups, were selected and are included to highlight essential aspects of their processes. Images are also incorporated to illustrate the students' work and examples of their actions in the class.

The students' thinking stages

After recognizing a pattern in the students' thinking, a descriptive account of the five performed stages was produced. The characterization of each thinking stage is presented below, and evidence of its content is provided.

Thinking about hand shape

The groups focused on the full-size images of their members' hands. They mostly talked about the shape of the human hand. They also discussed the security purpose of the recognition system. They realized that the hand differs from person to person, although they acknowledged that some people may have similar hands, which makes it difficult for the system to distinguish between hands of different individuals. In one group, the students mentioned that better biometric processes, such as fingerprints, may be used. They talked about forensic investigations, in which the recognition of individuals is largely done through fingerprints.

The students also observed the images of the hands next to their real hands. In most groups, each student concentrated on the image of her/his own hand. In some cases, the students drew the contour of the hand on the photocopy (some with pencil, others using a white corrector pen for better visualization over the dark photocopy).

Then, the students wondered about which biometrics they would use in their system. The several groups were focusing on elements of hand geometry that could be transformed into measures; at some point, they began to decide how to mathematically describe the hand shape.

Thinking about hand features

In the groups, much discussion was produced around the best features of the hand to choose for an adequate description of the hand. The discussion mainly involved locating specific points (e.g., the midpoint of the wrist line, the fingertips) or identifying straight segments (e.g., the segment from the base to the tip of each finger, the segments between the finger joints). During this period, the students talked about the anatomy of the hand. They referred to phalanges, joints, palm, and wrist. In some groups they used the finger names, in other groups they used numbers to designate the fingers. Various materials were used at this stage. All groups used rulers, and some also used drawing compasses (Figure 3). The groups also turned to the printed outline of the hand that was provided and drew sketches of the features in it.

Decisions about the features to consider were quite varied between groups. The number of features ranged from a minimum of 6 to a maximum of 20. The choices were also very diverse; some chose, for example, the lengths of the fingers and the widths of the phalanges measured at the midpoint of the finger length; others chose to consider the lengths of the phalanges and the width of the palm, and others used a circle centered in the middle of the palm and found its area (Figure 4). They integrated their knowledge about the hand's anatomy for deciding on the numerical inputs that would be used by the system.

Figure 3. Students making measurements on the photocopy of a hand.

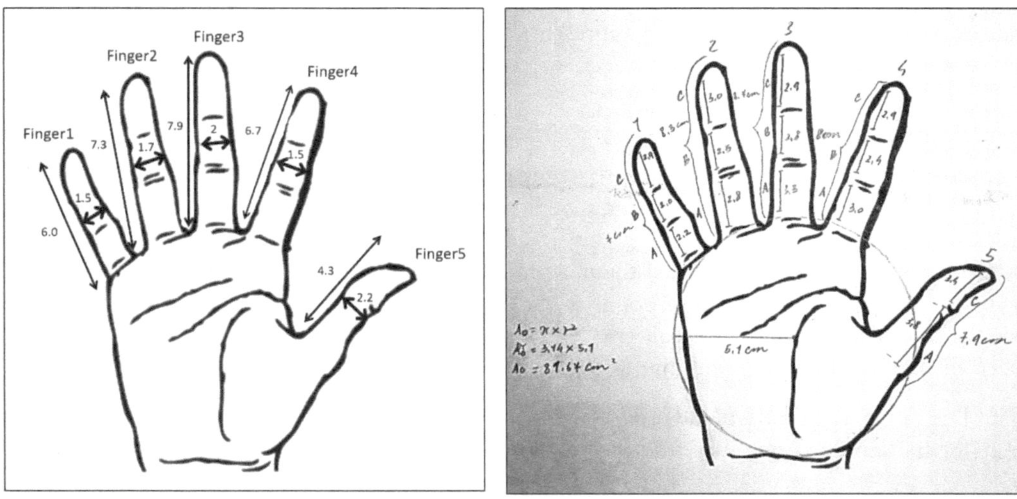

Figure 4. Examples of the features chosen by the groups (lengths, widths, and areas).

Thinking about data collection

In general, the groups tried to simplify the description of the hand by looking at how different features allowed sufficient differentiation between the hands of the group members. In addition, all features corresponded to measures (in centimeters or in millimeters) and were translated into numbers.

In Group A, the Student A3 helped his colleagues organize the recording of the measurements, using the hand contour sheet. He drew the reference lines they agreed to use as hand features.

Student A3: First the length. Then you mark the middle here. Then you measure (the width) from here to here.

Student A1: I get 7.9 (the middle finger).

Student A2: In mine, I get 8.2.

Student A3: Mine is 8.2 and the width is 1.9.

Student A1: Mine doesn't even reach 8. I'm the biggest but I'm the one with the smallest hand.

Student A2: On the thumb, I have 2 centimeters (of width).

Student A1: On the thumb, I have 1.9 centimeters.

The students A3 and A2 compared their hands, placing one against the other. They realized they were different in size, one of them slightly larger than the other, which seemed to make them confident about their data.

In some groups the concern with differentiation was greater, especially with students arguing about preventing the system from accepting an outsider's hand that could be too much like a member's hand. Therefore, the early concern with false positives revealed a clear connection with the security purpose of the system.

The groups created ways of organizing and representing the numerical data corresponding to the features they extracted from the hand. Most groups made a table to represent these data. In the table, for each element of the group, they wrote the values obtained for the various features (Figure 5). There were cases where they did not use a table but wrote the values obtained for each element of the group as ordered lists (like vectors).

At the end of this stage, the students performed the enrollment and the data acquisition process that the recognition system must ensure. The quantitative data they obtained correspond to indexed values, in the sense that each value refers to one of the registered users and one of the hand features. It is, therefore, a quantitative database that represents the identities of the members of the group. This database is a fundamental step in the process of mathematizing the problem.

Thinking about the decision model

The students began to reflect on how to compare hands of different individuals, based on the numerical data. They examined the values in their tables and looked at the range of variation of each feature.

In Group D, the students inspected the table in which they recorded 5 of the several features they extracted (each feature, *A* to *E*, corresponds to the sum of the finger widths in the joints). Student D5 circled the closest values along each column (Figure 6). In the last row, they also calculated the difference between the maximum and the minimum in each column, which gave them a sense of the variability.

Student D5: Teacher, we have made progress! We have close values [pointing in the table] on different fingers (for the same person). We are now looking at the same finger (in different persons).

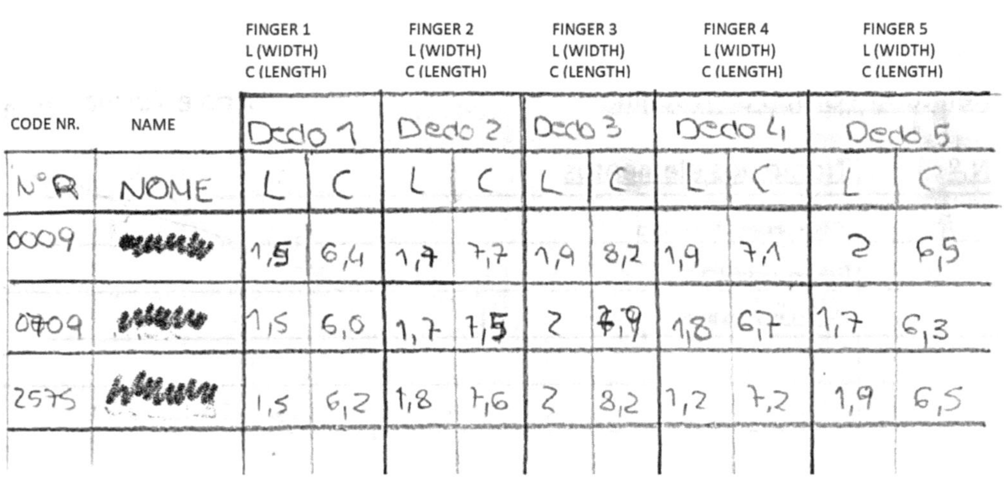

CODE NR.	NAME	FINGER 1 L (WIDTH) C (LENGTH)		FINGER 2 L (WIDTH) C (LENGTH)		FINGER 3 L (WIDTH) C (LENGTH)		FINGER 4 L (WIDTH) C (LENGTH)		FINGER 5 L (WIDTH) C (LENGTH)	
		Dedo 1		Dedo 2		Dedo 3		Dedo 4		Dedo 5	
Nº R	NOME	L	C	L	C	L	C	L	C	L	C
0009	[redacted]	1,5	6,4	1,7	7,7	1,9	8,2	1,9	7,1	2	6,5
0709	[redacted]	1,5	6,0	1,7	7,5	2	7,9	1,8	6,7	1,7	6,3
2575	[redacted]	1,5	6,2	1,8	7,6	2	8,2	1,2	7,2	1,9	6,5

Figure 5. Table with the biometric data of a group, showing the width (L) and the length (C) of the fingers, numbered from 1 to 5 (names crossed out to keep anonymity; translation added).

	A	B	C	D	E
	11,1	11,7	12,7	12,1	9,4
	9,3	13,7	13,7	13,2	9,4
	9,3	11	11,5	10,7	8,9
	13,2	13,10	14,2	14,1	10,1
	10,5	12,4	13.6	12,6	8,2
	2,9	2,1	2,7	3,4	1,9

Figure 6. A table with biometric data of a group, showing the features *A, B, C, D, E*, for each user, and the last row with the difference between the column maximum and minimum (names crossed out in the first column to keep anonymity).

Teacher: There seem to be some similarities here, but can you distinguish the hands?

Student D3: Let's see what the range in this finger is . . .

Student D2: Let's see where the measures diverge the most.

Student D3: Should we see that on every finger?

In the various groups, students started to discuss about a mathematical procedure to compare a new scanned hand with the hands of the registered users. This process presented significant difficulties in all groups. The discussions involved the idea of creating one or more comparison rules, but the students could not readily imagine what such a rule might look like.

Practically all groups assumed that the features of an unknown hand had to be compared with the corresponding features of each person in the database. This idea may be called exhaustive comparison. Some groups, especially those that had selected a larger number of attributes, tried more "condensed" forms of comparison, combining some features together, for example, adding or averaging. In all cases, however, students realized that the matching could not be seen as an exact matching. Hence, they thought of close values and considered the absolute value distance between real numbers. For deciding on the acceptable distance, students established the error under which the matching would be considered acceptable.

Finding an appropriate threshold was not a simple task for the students in the various groups.

Student C2: Let's say an error of 0.1 (centimeters). By default, or by excess?

Student C3: By default, and by excess.

Student C2: On all fingers, yes. This one is 7.5. In this case, it is between 7.4 and 7.6.

Student C3: It's hard to decide the maximum error . . .

By pressing their hands against the table and observing the extended size, the students got a sense of the dilation of the hand when the incoming data are acquired. One group defined the tolerance as 2 mm (Figure 7). They decided that an unknown individual should be accepted when there is an enrolled user such that the absolute distance obtained, for all the features, is less than 2 mm. Otherwise, the system should reject the individual as unauthorized.

The various models that emerged in the groups used the absolute value distance, although with some variations, both in the value of the established tolerance and in the systematic way of carrying out the comparison. Although no group has symbolically formalized its acceptance/rejection model, the general concept can be expressed by the rule, where an unknown hand x is compared to each of the enrolled hands y_j:

Acceptance: $|x_i - y_{ij}| \leq \delta$, for every i feature
Rejection: $|x_i - y_{ij}| > \delta$, for at least one i feature

Thinking about prototype validation

Many of the students' difficulties concerned the reliability of the process for comparing the data from an unknown hand with the stored data. The fact that students were designing a process of capturing biometric data, using real-sized images of their own hands, and noticing differences and similarities between human hands, made them struggle on how to integrate that knowledge into a recognition system. They were paying much attention to security, that is, to avoiding false positives. But they also realized that due to several factors, such as the pressure applied by the person on the device, the weight loss, the occasional swelling of the hand, etc., the incoming measurements of an enrolled user could be different from the user's recorded measurements.

Teacher: Can you explain to me how you will define an error to accept or reject an individual? Have you thought of some way to find the error?

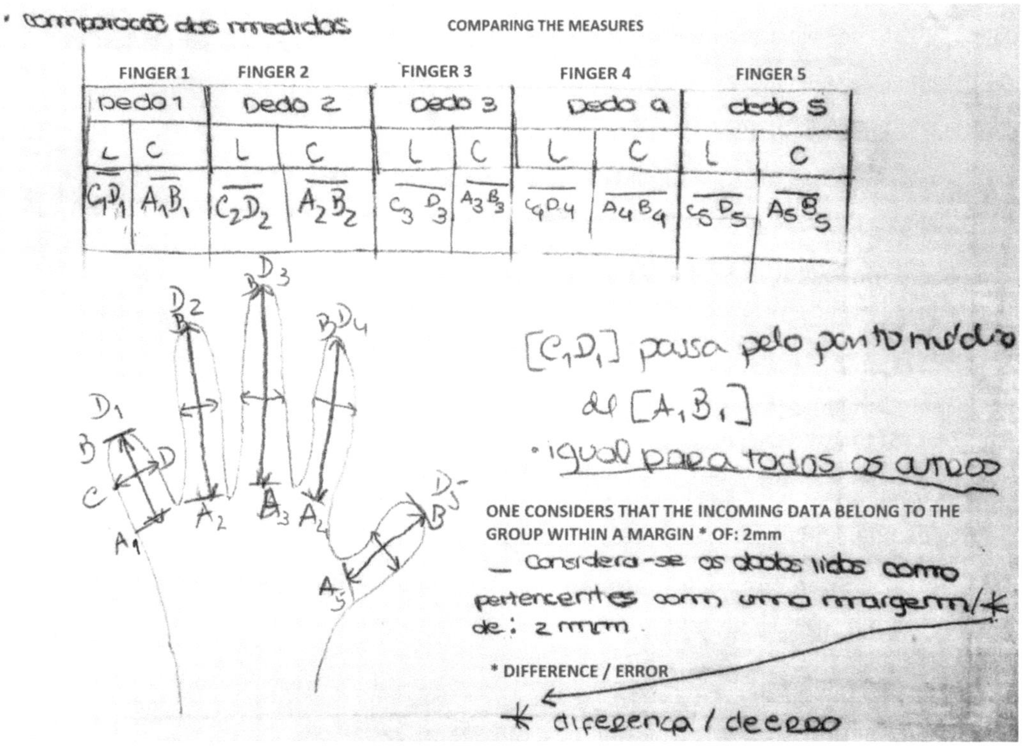

Figure 7. Student's notes explaining the comparison method using 10 features and the accepted error (translation added).

Student D5: If . . . If we don't accept any tolerance, then we run the risk of having a hand that should be accepted but falls outside the values we have . . . But if we accept a tolerance, then the (unknown) hand can be anyone's hand, so . . . I don't know . . . maybe not anyone's hand, but there may be three or four hands within that tolerance . . .

For testing the prototype designed by each group, other images from the second set of photocopies were randomly distributed. The students collected the necessary data from the unknown hand, according to the features considered in their biometric model. Then they checked whether the new test incoming hand belonged to the group or not. For the acceptance/rejection decision, they used the comparison model to evaluate the distance between the new captured features and the ones stored in the group's database.

After applying the comparison model, the groups accepted or rejected the "unknown hand" and validated the model produced for the recognition system. In one of the groups, the unknown hand was rejected, as the comparison model showed the error being exceeded for all members of the database.

Student A2: We must do the same as we did for our hands. In this case, we need to know all the values from this hand.

Student A1: Then we need to compare the values of the unknown hand against the ones of our hands, that is, in our table.

(. . .)

Teacher: So, this hand, was it rejected or accepted?

Student A2: It was rejected.

Teacher: Why?

Student A2: Because we compared the values, after setting an error. They don't match up with any member of our database. It exceeds the error in all cases.

This group validated their model and considered their prototype to be viable.

Another group, however, encountered a conflicting result. When they tested their prototype with a hand that belonged to one of the group members, the result they got was to reject that hand. Students were surprised by the result and looked for reasons to explain it. They started by thinking that wrong measures were taken of the test incoming hand. They carefully repeated the measurements but once again the result was to reject the hand. Finally, the students focused on their tolerance error and changed it so that the hand was accepted.

All groups finished their tests with satisfactory results, with some cases occurring where the tolerance had to be changed. Thus, the groups concluded the validation of their prototypes of a hand biometric recognition system.

Synthesis of results

As the results described above reveal, it was possible to discern a global pattern in the students' activity and thinking when faced with the design system problem.

The system involves a set of elements, including mathematical elements, and a set of relationships between them. The students in all groups paid attention to the purpose of the biometric recognition system, assuming that it was intended to recognize and authorize access to a pre-defined group of users. As a biometric system, its central characteristic is to relate biology (in this case, hand anatomy) and some metric associated with the geometry of the human hand. Finally, the problem requires the design of a system that takes (non-mathematical) inputs, transforms them through mathematical procedures and concepts, and generates (non-mathematical) outputs.

From a cognitive point of view, the interplay between mathematical modeling, especially the mathematization process, and the system design process was a prominent characteristic of student's thinking. That interplay, or the multi-dimensional nature of their thinking, is reflected in the way students incorporated equivalent concepts from mathematics and from the real situation. Two essential concepts fueled the students' reasoning during the construction of their prototypes. They concern the similarity and differentiation between hands: there is similarity between the hands of different individuals, which means uniformity of anatomical characteristics; there is variability between different hands and even between the hand of the same person if scanned at different times. Those concepts became apparent in their interactions with the hand images and their observation of the shape and size of the hands, also reinforced by the hand measurements. Although such concepts arose from their manipulation of real and tangible elements and from their extra-mathematical knowledge, they became key concepts within the mathematization process and led to measurement, data acquisition, deviation, and approximation. This suggests that the students were doing *mathematical modeling while designing* the system. This claim is elucidated by the diagram in Figure 8, which presents a generic representation of the processes carried out by the students, except for specific variations and particularities that were observed in the individual groups (as reported above).

In the diagram, the mathematical modeling process and the design process run together, and it becomes evident that the two processes mirror each other. The various phases of the biometric recognition system design process, namely enrollment, matching, prototyping, and validation, clearly emerge in the pattern found. At the same time, the crucial points in the construction of the mathematical model that is responsible for the system's output (decision to accept or reject an incoming hand) are quite visible. The model is based on the mathematical concepts of distance and error. Finally, the prototype testing entails the validation of the mathematical model and its subsequent revision in case the results are not in accordance with the real situation.

The results achieved are positive regarding the potential of integrated STEM contexts in which the system design process is a driver of the students' mathematical modeling activity. Other studies had already highlighted similar potentialities (English, 2019; Hallström & Schönborn, 2019; Kelley & Knowles, 2016; Leung, 2019). The dominant feature of a system design is that it requires a comprehensive and holistic understanding of a real-world problem, leading the modeler to think about the various dimensions of the problem, including its mathematical components, and to deal with them in an integrated and multi-dimensional way (Almeida, 2018; Borromeo Ferri & Mousoulides, 2017; Brown & Edwards, 2011). As other researchers have concluded, there are multiple opportunities for students learning while designing (Bock et al., 2019; English, 2019; English et al., 2017). From our analysis it can also be inferred that there are multiple opportunities for students mathematizing while designing.

This emerges as a relevant result regarding students' mathematical thinking in a STEM context where science, engineering and mathematics were integrated. While the students went through a simplified prototyping cycle, within the reach of their abilities and knowledge, they went through the ideal cycle of mathematical modeling (Blum & Leiβ, 2007). Students understood the problem, especially realizing the need to simultaneously deal with the uniformity and differentiability of hand geometry. They simplified the problem by making decisions about the measurements and features to extract as to meet those conditions. They organized and systematized their data analogously to representing matrices or sets of vectors. Finally, students worked mathematically to create an algorithmic procedure that resulted in the decision of admitting or declining access to an individual. They ended up testing and validating their models by reformulating aspects such as the acceptable error or the number of features to include in the model.

From this point of view, the ideal modeling cycle proves to be a robust theoretical construct for describing and synthesizing the students' cognitive processes in a design problem. Moreover, it should be emphasized that the modeling cycle is mirrored by a design cycle that runs in parallel. Therefore, students' mathematical thinking develops in close connection with system thinking.

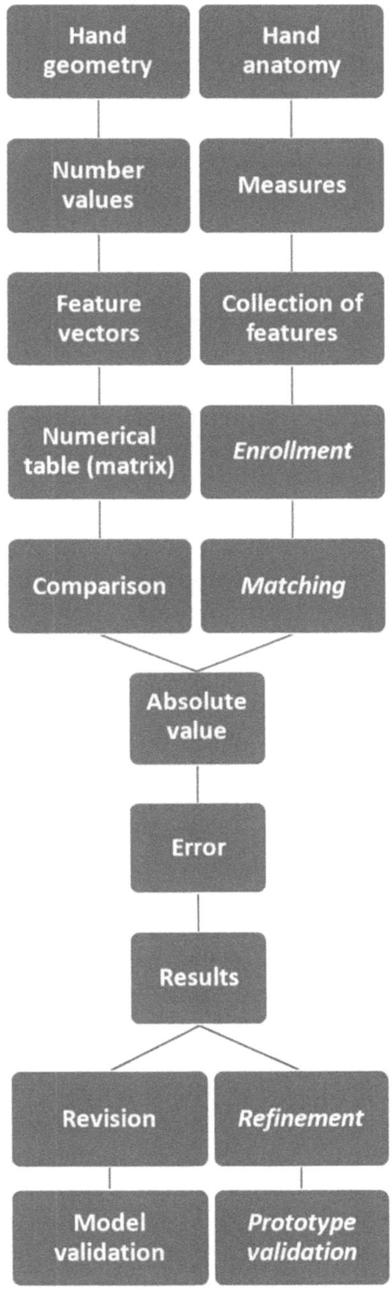

Figure 8. A representation of the global pattern of the students' modeling process while designing the system.

System thinking pushes and drives students' mathematical thinking, from the moment of identifying the system requirements to the validation of a prototype. This further shows that one quality of the system is highly relevant: it is a system in which mathematics underlies fundamental aspects of its functioning. In that case, the relation between the model and the prototype becomes very strong (as illustrated in Figure 8), such that they bond to each other (Birta & Arbez, 2007; Isa & Liem, 2014).

Conclusion

In this study we have undertaken a cognitive perspective, aiming to understand how students' mathematical thinking is activated and nurtured in solving a design problem, in the context of creating a biometric system for identity verification. Drawing on data collected during two design-based research cycles with 9th grade classes, we carried out a directed qualitative analysis, theoretically informed by the structure and sub-processes of the mathematical modeling cycle (Blum & Leiβ, 2007) and the prototyping cycle (Laudon & Laudon, 2014).

Rather than focusing on student's particular modeling routes, we have concentrated on the global pattern of student's thinking while designing a system. The global pattern revealed a strong interplay between the design process and the mathematical modeling process (Isa & Liem, 2014; Leung, 2019). This provides evidence that this challenging design problem represents an important boundary crossing opportunity. The research carried out so far offers clear indicators that integrated STEM contexts enable and favor the boundary crossing, helping students to develop relevant concepts and knowledge, both in mathematics and in other connected disciplines (English, 2016, 2019; English et al., 2017; Kertil & Gurel, 2016).

The need to achieve a comprehensive and multi-dimensional understanding of the problem stimulates students' thinking processes of a multi-dimensional nature (Araya Schulz, 2016; Borromeo Ferri & Mousoulides, 2017). This holistic understanding brings with it a positive blend between mathematical and extra-mathematical knowledge. Thus, the activation of diverse knowledge bases played a crucial role in students' cognitive processes, as other studies have also suggested and supported (Brown & Edwards, 2011; Czocher, 2016; Stillman, 2000). In the students' modeling processes, we observed an integrative approach to the problem. Students intensely discussed and analyzed aspects of the anatomy of the human hand, its geometry, and the ways to describe it in mathematical terms. Likewise, they vigorously debated and struggled to formulate a mathematical model that would allow generation of an output from the biometric system. They resorted to the concepts of absolute value distance and error (or tolerance) that were part of their mathematical knowledge base. This indicates a sensible reasoning, given that in actual biometric recognition systems such concepts are fundamental to their functioning (cf. Duta, 2009).

Kertil and Gurel (2016) have already stated that: "all STEM activities are not modeling activities, but students may experience the mathematical modeling process while working on many of them" (p. 46). From what we observed in our study, we can assert that a system design activity may be seen as a mathematical modeling activity. In particular, a substantial overlapping between the mathematization process and the design process stood out from our results. In fact, from the analysis of students' mathematical thinking we may infer that the mathematization process was anticipated (Almeida, 2018; Niss, 2010) in line with the anticipation of the system's operation.

The overlapping and interplay between mathematical modeling and system design were therefore prominent conclusions drawn from our analysis. This is related to various aspects of the task setting, such as putting into action practices from different areas, namely engineering design practices and mathematical practices and reasoning. The cyclical and iterative character of both the modeling and the design processes also strengthens the affinity between them. It was apparent that students engaged in reviewing their models and designs, which is highly significant since several studies have found that students and teachers often neglect the phase of revising and validating their models (Blum, 2015; Blum & Leiβ, 2007; Galbraith & Stillman, 2006; Zeytun et al., 2017). Another attribute of design problems is that they heavily rely on experimentation, manipulation, and simulation (often resorting to technological tools and software). As we have already reported in previous studies (Carreira & Baioa, 2011, 2018), experimenting and using physical materials gives students possibilities to deepen their understanding of a real problem, as they have the means to explore, investigate and inquire about tangible situations and systems (Diefes-Dux et al., 2008; English, 2019; Stohlmann et al., 2012). The several experiments and explorations made by the students in collecting hand features and examining those data (by resorting to images and to their own hands) strengthened their belief that a mathematical model of the human hand could be achieved to make the biometric recognition system work. This idea is corroborated by experts' assertions, such as: "experiments show that the physical dimensions of a human hand contain information that is capable to verify the identity of an individual" (Varchol & Levický, 2007, p. 82). In fact, experimentation plays a key role in system thinking. At the same time, it is important to stress that students were working on purposeful prototyping (Petrakis et al., 2021). This means that understanding the prototype's purpose prior to initiating the prototyping process is decisive. In our data we found evidence that students developed the design and mathematical modeling processes by taking in the purpose of the biometric recognition system they were asked to develop.

Finally, some limitations of the study should be mentioned. A clear absence in the students' activity was the use of technological tools, except for the calculator, which they used at various times for simple calculations. The use of technology, such as the spreadsheet, would be an appropriate and enriching opportunity for solving the task. This option, although it was considered, especially in the 2nd research cycle, could not be realized. This was due to difficulties in managing the available class time and to the lack of students' experience with the spreadsheet that would demand additional conditions and time. However, as this is a key area in STEM integrated contexts, we anticipate that technology will be endorsed and purposefully integrated in our future research involving system design activities. Furthermore, the task presupposed a very simplified and rudimentary version of a recognition system based on hand biometrics. Each group in the class worked with a very small number of enrolled users and had very few image samples per user. This has not favored a deeper analysis of the differences between several samples of the same hand, which could motivate the use of relevant statistical concepts, such as mean and standard deviation. Therefore, statistical concepts and notions might also be appropriate mathematical tools in more complex versions of the design problem of a hand biometrics recognition system.

Acknowledgments

We wish to express our gratitude to the anonymous Reviewers and to the Editors for the thoughtful and constructive suggestions and feedback on our work.

Disclosure statement

No potential conflict of interest was reported by the author(s).

ORCID

Ana Margarida Baioa http://orcid.org/0000-0002-6620-6031
Susana Carreira http://orcid.org/0000-0002-0746-7258

References

Albarracín, L., Ärlebäck, J., Civil, E., & Gorgorio, N. (2019). Extending modelling activity diagrams as a tool to characterize mathematical modelling processes. *The Mathematics Enthusiast*, *16*(1), 10. https://scholarworks.umt.edu/tme/vol16/iss1/10

Almeida, L. M. W. (2018). Considerations on the use of mathematics in modeling activities. *ZDM Mathematics Education*, *50*(1–2), 19–30. https://doi.org/10.1007/s11858-017-0902-4

Araya Schulz, R. (2016). STEM y modeliamento matemático. *Cuadernos de Investigación Y Formación En Educación Matemática*, *11*(15) , 291–317. https://revistas.ucr.ac.cr/index.php/cifem/article/view/23838/24006

Ärlebäck, J. (2009). On the use of realistic Fermi problems for introducing mathematical modelling in school. *The Mathematics Enthusiast*, *6*(3), 4. https://scholarworks.umt.edu/tme/vol6/iss3/4

Aslaksen, E. (2013). *The system concept and its application to engineering*. Springer.

Baioa, A. M., & Carreira, S. (2018). Tarefas de modelação matemática em contexto STEM: Um possível recurso para a aula de matemática. In A. Rodrigues, A. Barbosa, A. Santiago, A. Domingos, C. Carvalho, C. Ventura, C. Costa, H. Rocha, J. M. Matos, L. Serrazina, M. Almeida, P. Teixeira, R. Carvalho, R. Machado, and S. Carreira (Eds.), *Livro de Atas do EIEM 2018 - Encontro de Investigação em Educação Matemática* (pp. 113–129). SPIEM, Coimbra, Portugal. http://spiem.pt/DOCS/ATAS_ENCONTROS/atas_EIEM_2018.pdf

Baioa, A. M., & Carreira, S. (2019b). Tarefas de modelação matemática com recurso a simulações em contexto STEM: Ciclo de modelação e competências no 9.º ano. In N. Amado, A. P. Canavarro, S. Carreira, R. T. Ferreira, and I. Vale (Eds.), *Livro de Atas do EIEM 2019 - Encontro de Investigação em Educação Matemática* (pp. 73–90). SPIEM, Alte-Loulé, Portugal. http://spiem.pt/DOCS/ATAS_ENCONTROS/atas_EIEM_2019.pdf

Baioa, A. M., & Carreira, S. (2019a). Simulations and prototypes in mathematical modelling. In U. T. Jankvist, M. Van Den Heuvel-
panhuizen, and M. Veldhuis (Eds.), *Proceedings of the Eleventh Congress of the European Society for Research in Mathematics Education - CERME 11* (pp. 1136–1143). Utrecht University and ERME, Utrecht, The Netherlands. https://hal.archives-ouvertes.fr/hal-02408688

Birta, L. G., & Arbez, G. (2007). *Modelling and simulation – Exploring dynamic system behaviour*. Springer.

Blomhøj, M., & Jensen, T. H. (2003). Developing mathematical modelling competence: Conceptual clarification and educational planning. *Teaching Mathematics and Its Applications*, *22*(3), 123–139. https://doi.org/10.1093/teamat/22.3.123

Blum, W., & Borromeo Ferri, R. (2009). Mathematical modelling: Can it be taught and learnt? *Journal of Mathematical Modelling and Application*, *1*(1), 45–58. https://proxy.furb.br/ojs/index.php/modelling/article/view/1620/1087

Blum, W., & Leiβ, D. (2007). How do students and teachers deal with modelling problems? In C. Haines, P. Galbraith, W. Blum, & S. Khan (Eds.), *Mathematical modelling: Education, engineering and economics - ICTMA 12* (pp. 222–231). Woodhead Publishing. https://doi.org/10.1533/9780857099419.5.221

Blum, W. (2015). Quality teaching of mathematical modelling: What do we know, what can we do? In S. J. Cho (Ed.), *Proceedings of the 12th International Congress on Mathematical Education – Intellectual and Attitudinal Challenges* (pp. 73–96). Seoul, South Korea: Springer. https://doi.org/10.1007/978-3-319-12688-3_9

Bock, W., Bracke, M., & Capraro, P. (2019). Mathematical modeling of musical fountains and light organs – Where is the M in interdisciplinary STEM projects? In U. T. Jankvist, M. Van Den Heuvel-panhuizen, and M. Veldhuis (Eds.), *Proceedings of the Eleventh Congress of the European Society for Research in Mathematics Education - CERME 11* (pp. 4739–4746). Utrecht University and ERME, Utrecht, The Netherlands. https://hal.archives-ouvertes.fr/hal-02410321

Borromeo Ferri, R. (2007a). Personal experiences and extra-mathematical knowledge as an influence factor on modelling routes of pupils. In D. Pitta-Pantazi, and G. Philippou (Eds.), *Proceedings of the 5th Congress of the European Society for Research in Mathematics Education - CERME 5* (pp. 2080–2089). University of Cyprus and ERME, Larnaca, Cyprus. http://erme.site/wp-content/uploads/CERME5/WG13.pdf

Borromeo Ferri, R., & Mousoulides, N. (2017). Mathematical modelling as a prototype for interdisciplinary mathematics education? – Theoretical reflections. In T. Dooley, and G. Gueudet (Eds.), *Proceedings of the Tenth Congress of the European Society for Research in Mathematics Education - CERME 10* (pp. 899–907). DCU Institute of Education & ERME, Dublin, Ireland. https://hal.archives-ouvertes.fr/hal-01933490

Borromeo Ferri, R. (2006). Theoretical and empirical differentiations of phases in the modelling process. *ZDM Mathematics Education*, *38*(2), 86–95. https://doi.org/10.1007/BF02655883
Borromeo Ferri, R. (2018). *Learning how to teach mathematical modelling in school and teacher education.* Springer. https://doi.org/10.1007/978-3-319-68072-9
Borromeo Ferri, R. (2007b). Modelling problems from a cognitive perspective. In C. Haines, P. Galbraith, W. Blum, & S. Khan (Eds.), *Mathematical modelling: Education, engineering and economics - ICTMA 12* (pp. 260–270). Woodhead Publishing. https://doi.org/10.1533/9780857099419.5.260
Breiner, J., Harkness, M., Johnson, C. C., & Koehler, C. (2012). What is STEM? A discussion about conceptions of STEM in education and partnerships. *School Science and Mathematics*, *112*(1), 3–11. https://doi.org/10.1111/j.1949-8594.2011.00109.x
Brown, J. P., & Edwards, I. (2011). Modelling tasks: Insight into mathematical understanding. In G. Kaiser, W. Blum, R. Borromeo Ferri, & G. Stillman (Eds.), *Trends in teaching and learning of mathematical modelling* (pp. 187–197). Springer. https://doi.org/10.1007/978-94-007-0910-2_20
Busse, A. (2005). Individual ways of dealing with the context of realistic tasks – First steps towards a typology. *ZDM Mathematics Education*, *37*(5), 354–360. https://doi.org/10.1007/s11858-005-0023-3
Carreira, S., Baioa, A. M., & Almeida, M. L. W. (2020). Mathematical models and meanings by school and university students in a modelling task. *Avances de Investigación En Educación Matemática*, *17*(17), 67–83. https://doi.org/10.35763/aiem.v0i17.308
Carreira, S., & Baioa, A. M. (2018). Mathematical modelling with hands-on experimental tasks: On the student's sense of credibility. *ZDM Mathematics Education*, *50*(1), 201–215. https://doi.org/10.1007/s11858-017-0905-1
Carreira, S., & Baioa, A. M. (2011). Students' modelling routes in the context of object manipulation and experimentation in mathematics. In G. Kaiser, W. Blum, R. Borromeo Ferri, & G. Stillman (Eds.), *Trends in teaching and learning of mathematical modelling* (pp. 211–220). Springer. https://doi.org/10.1007/978-94-007-0910-2_22
Cobb, P. (2000). Conducting teaching experiments in collaboration with teachers. In A. E. Kelly & R. A. Lesh (Eds.), *Handbook of research design in mathematics and science education* (pp. 307–330). Lawrence Erlbaum Associates.
Czocher, J. A. (2016). Introducing modeling transition diagrams as a tool to connect mathematical modeling to mathematical thinking. *Mathematical Thinking and Learning*, *18*(2), 77–106. https://doi.org/10.1080/10986065.2016.1148530
Diefes-Dux, H. A., Hjalmarson, M. A., Miller, T. K., & Lesh, R. (2008). Model-eliciting activities for engineering education. In J. S. Zawojewski, H. A. Diefes-Dux, & K. J. Bowman (Eds.), *Models and modeling in engineering education: Designing experiences for all students* (pp. 17–35). Sense.
Duta, N. (2009). Hand shape. In S. Z. Li & A. Jain (Eds.), *Encyclopedia of biometrics* (2009 ed., pp. 682–687). Springer.
Dym, C. L., Agogino, A. M., Eris, O., Frey, D. D., & Leifer, L. J. (2005). Engineering design thinking, teaching, and learning. *Journal of Engineering Education*, *94*(1), 103–120. https://doi.org/10.1002/j.2168-9830.2005.tb00832.x
Dym, C. L. (2004). *Principles of mathematical modeling* (2nd ed.). Elsevier.
English, L. (2016). STEM education K-12: Perspectives on integration. *International Journal of STEM Education*, *3*(1), 3. https://doi.org/10.1186/s40594-016-0036-1
English, L., Ärlebäck, J., & Mousoulides, N. (2016). Reflections on progress in mathematical modelling research. In Á. Gutiérrez, G. C. Leder, & P. Boero (Eds.), *The second handbook of research on the psychology of mathematics education: The journey continues* (pp. 383–413). Sense Publishers.
English, L. (2015). STEM: Challenges and opportunities for mathematics education. In K. Beswick, T. Muir, and J. Wells (Eds.), *Proceedings of the 39th Conference of the International Group for the Psychology of Mathematics Education* (Vol. 1, pp. 3–18). PME - The International Group for the Psychology of Mathematics Education, Tasmania, Australia.
English, L., King, D., & Smeed, J. (2017). Advancing integrated STEM learning through engineering design: Sixth-grade students' design and construction of earthquake resistant buildings. *The Journal of Educational Research*, *110*(3), 255–271. https://doi.org/10.1080/00220671.2016.1264053
English, L. (2009). Promoting interdisciplinarity through mathematical modelling. *ZDM Mathematics Education*, *41*(1–2), 161–181. https://doi.org/10.1007/s11858-008-0106-z
English, L. (2019). Learning while designing in a fourth-grade integrated STEM problem. *International Journal of Technology and Design Education*, *29*(5), 1011–1032. https://doi.org/10.1007/s10798-018-9482-z
Fatima, F., Javed, M., Amjad, F., & Khan, U. G. (2018). An approach to enhance quality of the RAD model using agents. *Journal of American Science*, *14*(9), 47–55. https://doi.org/10.7537/marsjas140918.08
Galbraith, P., & Stillman, G. (2006). A framework for identifying student blockages during transitions in the modelling process. *ZDM Mathematics Education*, *38*(2), 143–162. https://doi.org/10.1007/BF02655886
Gallegos, R. (2018). Building bridges between the math education and the engineering education communities: A dialogue through modelling and simulation. In G. Kaiser, H. Forgasz, M. Graven, A. Kuzniak, E. Simmt, and B. Xu (Eds.), *Invited lectures from the 13th International Congress on Mathematical Education, ICME 13 Monographs* (pp. 501–519). Hamburg, Germany: Springer. https://doi.org/10.1007/978-3-319-72170-5_28

Haines, C., & Crouch, J. (2010). Remarks on a modelling cycle and interpreting behaviours. In R. Lesh, P. Galbraith, C. Haines, & A. Hurford (Eds.), *Modelling students' mathematical modelling competencies - ICTMA 13* (pp. 145–154). Springer.

Hallström, J., & Schönborn, K. J. (2019). Models and modelling for authentic STEM education: Reinforcing the argument. *International Journal of STEM Education*, *6*(1), 22. https://doi.org/10.1186/s40594-019-0178-z

Hsieh, H.-F., & Shannon, S. E. (2005). Three approaches to qualitative content analysis. *Qualitative Health Research*, *15* (9), 1277–1288. https://doi.org/10.1177/1049732305276687

Hubalovsky, S. (2010). Modelling of real kinematics situation as a method of the system approach to the algorithm development thinking. *International Journal of Applied Mathematics and Informatics*, *4*(4), 83–91. https://www.naun.org/main/UPress/ami/19-698.pdf

Isa, S. S., & Liem, A. (2014). Classifying physical models and prototypes in the design process: A study on the economical and usability impact of adopting models and prototypes in the design process. In D. Marjanovic, M. Storga, N. Pavkovic, and N. Bojcetic (Eds.), *Proceedings of the DESIGN 2014-13th International Design Conference* (pp. 2071–2082). Cavtat , Croatia: The Design Society. https://www.designsociety.org/publication/35348/

Kaiser, G., & Sriraman, B. (2006). A global survey of international perspectives on modelling in mathematics education. *ZDM Mathematics Education*, *38*(3), 302–310. https://doi.org/10.1007/BF02652813

Kelley, T. R., & Knowles, J. G. (2016). A conceptual framework for integrated STEM Education. *International Journal of STEM Education*, *3*(1), 11. https://doi.org/10.1186/s40594-016-0046-z

Kelly, A., Baek, J., Lesh, R., & Bannan-Ritland, B. (2008). Enabling innovations in education and systematizing their impact. In A. Kelly, R. Lesh, & J. Baek (Eds.), *Handbook of design research methods in education: Innovations in science, technology, engineering, and mathematics learning and teaching* (pp. 3–18). Routledge.

Kertil, M., & Gurel, C. (2016). Mathematical modelling: A bridge to STEM education. *International Journal of Education in Mathematics, Science and Technology*, *4*(1), 44–55.

Knuutila, T. (2005). Models, representations, and mediation. *Philosophy of Science*, *72*(5), 1260–1271. https://doi.org/10.1086/508124

Laudon, K., & Laudon, J. (2014). *Management information systems* (13th ed.). Pearson Education.

Leung, A. (2019). Exploring STEM pedagogy in the mathematics classroom: A tool-based experiment lesson on estimation. *International Journal of Science and Mathematics Education*, *17*(7), 1339–1358. https://doi.org/10.1007/s10763-018-9924-9

Li, S. Z., & Jain, A. (Eds.). (2009). *Encyclopedia of biometrics* (2009 ed.). Springer. https://doi.org/10.1007/978-0-387-73003-5

Michelsen, C. (2006). Functions: A modelling tool in mathematics and science. *ZDM Mathematics Education*, *38*(3), 269–280. https://doi.org/10.1007/BF02652810

Michelsen, C. (2015). Mathematical modeling is also physics—interdisciplinary teaching between mathematics and physics in Danish upper secondary education. *Physics Education*, *50*(4), 489–494. https://doi.org/10.1088/0031-9120/50/4/489

Moore, T., & Smith, K. (2014). Advancing the state of the art of STEM integration. *The Journal of STEM Education: Innovations and Research*, *15*(1), 5–10. https://www.jstem.org/jstem/index.php/JSTEM/article/view/1917/1581

Mousoulides, N., & English, L. (2008). Modelling with data in Cypriot and Australian classrooms. In O. Figueras, J. L. Cortina, S. Alatorre, T. Rojano, and A. Sepulveda (Eds.), *Proceedings of the 32nd International Conference of the International Group for the Psychology of Mathematics Education* (Vol. 3, pp. 423–430). PME - The International Group for the Psychology of Mathematics Education, Morelia, Mexico.

Mousoulides, N., & English, L. (2011). Engineering model eliciting activities for elementary school students. In G. Kaiser, W. Blum, R. B. Ferri, & G. Stillman (Eds.), *Trends in teaching and learning of mathematical modelling - ICTMA 14* (pp. 221–230). Springer.

Niss, M. (2010). Modeling a crucial aspect of students' mathematical modeling. In R. Lesh, P. Galbraith, C. Haines, & A. Hurford (Eds.), *Modelling students' mathematical modelling competencies - ICTMA 13* (pp. 43–59). Springer.

OECD. (2018). *PISA 2022 mathematics framework (Draft)*. https://pisa2022-maths.oecd.org/files/PISA%202022%20Mathematics%20Framework%20Draft.pdf

Petrakis, K., Wodehouse, A., & Hird, A. (2021). Physical prototyping rationale in design student projects: An analysis based on the concept of purposeful prototyping. *Design Science*, *7*, E7. https://doi.org/10.1017/dsj.2021.6

Potter, W. J., & Levine-Donnerstein, D. (1999). Rethinking validity and reliability in content analysis. *Journal of Applied Communication Research*, *27*(3), 258–284. https://doi.org/10.1080/00909889909365539

Ramírez-Montes, G., Henriques, A., & Carreira, S. (2021). Undergraduate students' learning of linear algebra through mathematical modelling routes. *Canadian Journal of Science, Mathematics and Technology Education*. Advance online publication. https://doi.org/10.1007/s42330-021-00149-3

Salado, A., Chowdhury, A. H., & Norton, A. (2019). Systems thinking and mathematical problem solving. *School Science and Mathematics*, *119*(1), 49–58. https://doi.org/10.1111/ssm.12312

Schukajlow, S., Kaiser, G., & Stillman, G. (2018). Empirical research on teaching and learning of mathematical modelling: A survey on the current state-of-the-art. *ZDM Mathematics Education*, *50*(1–2), 5–18. https://doi.org/10.1007/s11858-018-0933-5

Stillman, G. (2000). Impact of prior knowledge of task context on approaches to applications tasks. *Journal of Mathematical Behavior*, *19*(1), 333–361. https://doi.org/10.1016/S0732-3123(00)00049-3

Stohlmann, M., Moore, T. J., & Roehrig, G. H. (2012). Considerations for teaching integrated STEM Education. *Journal of Pre-College Engineering Education Research*, *2*(1), 4. https://doi.org/10.5703/1288284314653

Tezer, M. (2020). The role of mathematical modelling in STEM integration and education. In K. G. Fomunyam (Ed.), *Theorizing STEM education in the 21st century*. IntechOpen. https://www.intechopen.com/chapters/68547

Tytler, R., Williams, G., Hobbs, L., & Anderson, J. (2019). Challenges and opportunities for a STEM interdisciplinary agenda. In B. Doig, J. Williams, D. Swanson, R. Borromeo Ferri, & P. Drake (Eds.), *Interdisciplinary mathematics education - ICME-13 monographs* (pp. 51–81). Springer. https://doi.org/10.1007/978-3-030-11066-6_5

Varchol, P., & Levický, D. (2007). Using of hand geometry in biometric security systems. *Radioengineering Journal*, *16* (4), 82–87. http://www.radioeng.cz/fulltexts/2007/07_04_082_087.pdf

Zeytun, A., Cetinkaya, B., & Erbas, A. (2017). Understanding prospective teachers' mathematical modeling processes in the context of a mathematical modeling course. *EURASIA Journal of Mathematics, Science and Technology Education*, *13*(3), 691–722. https://doi.org/10.12973/eurasia.2017.00639a

Appendix 1

Hand Recognition System Task

Hand geometry is often used in security systems and is one of the most used worldwide. **Biometry** based on hand geometry uses a set of measurements that can be extracted, by considering some distinctive points of the hand, the five fingertips and the four valleys between them.

In this task, your group will create a database and a recognition system, using hand geometry, which may eventually be employed in the school, for example, to access the cafeteria or a specific computer room.

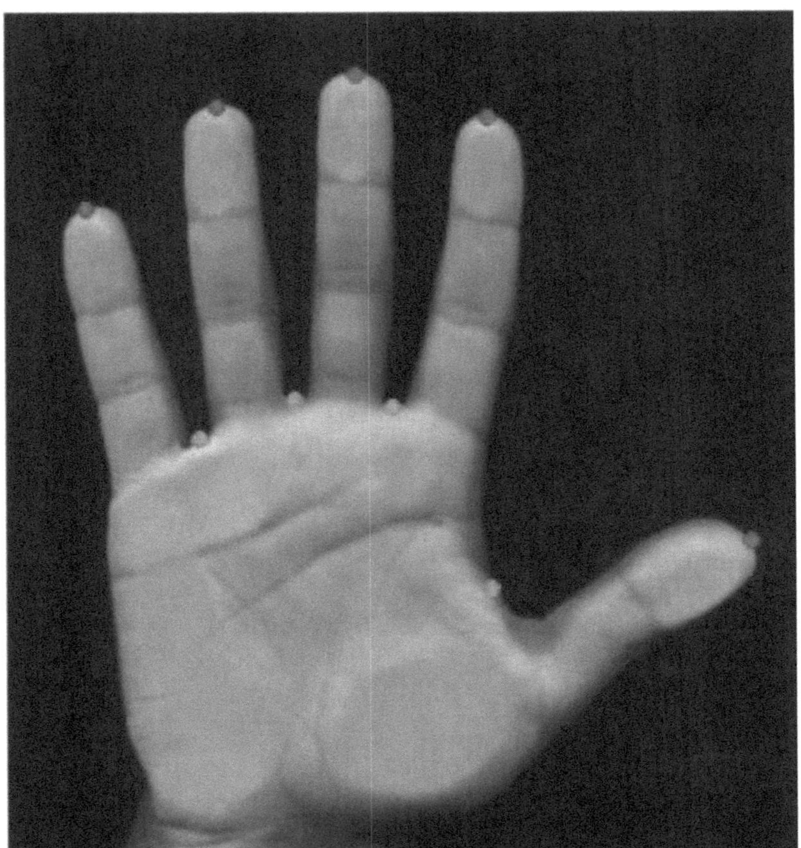

Image retrieved from: http://bip.inesctec.pt/113/noticia07.html

From the experience . . .

- You have a set of images (photocopies) of your palms at your disposal
- Decide what measures to collect from your hands

- Create a database with hand measurements
- You can use a recording sheet for each hand, like the given one.

. . . To the model

- Develop a mathematical model that allows:
  - accepting an authorized incoming hand
  - rejecting an unauthorized incoming hand
- Test your system with a hand drawn at random from the class and a hand drawn at random from your group.

Guidelines for preparing the report:

- Your understanding of the initial situation
- Assumptions about the measures to be used, explaining your options
- The experiments carried out and the data collected
- The mathematical model built for the verification system
- The validation of the model and the results obtained
- Difficulties experienced during the activity
- Assessment of the task in relation to motivating level, relation to reality and other disciplines, and mathematical concepts involved.

Index

Note: **Bold** page numbers refer to tables; *italic* page numbers refer to figures and page numbers followed by "n" denote endnotes.

Kolibri 360 GmbH
Ferdinand-Jühlke-Straße 7
99095 Erfurt, Deutschland
produktsicherheit@kolibri360.de